Carbohydrate Antigens

A C S   S Y M P O S I U M   S E R I E S   **519**

Carbohydrate Antigens

Per J. Garegg, EDITOR
University of Stockholm

Alf A. Lindberg, EDITOR
Karolinska Institute

Developed from a symposium sponsored
by the Division of Carbohydrate Chemistry
of the American Chemical Society
at the Fourth Chemical Congress of North America
(202nd National Meeting of the American Chemical Society)
New York, New York,
August 25–30, 1991

American Chemical Society, Washington, DC 1993

Library of Congress Cataloging-in-Publication Data

Carbohydrate antigens /   Per J. Garegg, editor, Alf A. Lindberg, editor.

   p.   cm.—(ACS symposium series, ISSN 0097–6156; 519)

"Developed from a symposium sponsored by the Division of Carbohydrate Chemistry of the American Chemical Society at the Fourth Chemical Congress of North America (202nd National Meeting of the American Chemical Society), New York, New York, August 25–30, 1991."

Includes bibliographical references and indexes.

ISBN 0–8412–2531–1

1. Carbohydrates—Immunology—Congresses.

I. Garegg, Per J. II. Lindberg, Alf A. III. American Chemical Society. Division of Carbohydrate Chemistry. IV. American Chemical Society. Meeting (202nd: 1991: New York, N.Y.) V. Series.

QR186.6.C37C37   1993
616.07′92—dc20   92–39537
                                    CIP

The paper used in this publication meets the minimum requirements of American National Standard for Information Sciences—Permanence of Paper for Printed Library Materials, ANSI Z39.48–1984.   ∞

Foreword

THE ACS SYMPOSIUM SERIES was first published in 1974 to provide a mechanism for publishing symposia quickly in book form. The purpose of this series is to publish comprehensive books developed from symposia, which are usually "snapshots in time" of the current research being done on a topic, plus some review material on the topic. For this reason, it is necessary that the papers be published as quickly as possible.

Before a symposium-based book is put under contract, the proposed table of contents is reviewed for appropriateness to the topic and for comprehensiveness of the collection. Some papers are excluded at this point, and others are added to round out the scope of the volume. In addition, a draft of each paper is peer-reviewed prior to final acceptance or rejection. This anonymous review process is supervised by the organizer(s) of the symposium, who become the editor(s) of the book. The authors then revise their papers according to the recommendations of both the reviewers and the editors, prepare camera-ready copy, and submit the final papers to the editors, who check that all necessary revisions have been made.

As a rule, only original research papers and original review papers are included in the volumes. Verbatim reproductions of previously published papers are not accepted.

M. Joan Comstock
Series Editor

This volume is dedicated to the memory
of Michael Heidelberger.

Acknowledgment

The Carbohydrate Division of the American Chemical Society gratefully acknowledges financial support to the "Carbohydrate Antigens" Symposium from the following companies:

Carlsberg A/S

Eli Lilly and Company

Glycomed Inc.

Hoffmann—La Roche Inc.

Pfanstiehl Laboratories

Schering—Plough Research

Contents

INDEXES

Preface

IN 1923, MICHAEL HEIDELBERGER AND OSWALD T. AVERY showed that the antigenic part of the capsule of the bacterium *Streptococcus pneumoniae* was a polysaccharide, not, as previously thought, a protein. It became clear that carbohydrate structures (oligosaccharides) are indeed involved in a myriad of cell interactions and biological regulatory processes, usually as conjugates (for example, bound to proteins in glycoproteins or to lipids in glycolipids). Saccharides had at first been considered materials for structure and energy storage or primary metabolites that arose from photosynthesis and were destined for further metabolic transformation. After Heidelberger and Avery's work, saccharides emerged as key substances in other biological processes. Indeed, 1923 might be considered the year of birth of what is now termed glycobiology.

Recent progress in biomedical research and in synthetic and structural carbohydrate chemistry has confirmed that oligo- and polysaccharide structures are involved in a host of biological mechanisms. Rapid progress in immunology and in structural chemistry has revealed the molecular basis for viral and bacterial recognition of and adhesion to specific cell structures. Detailed studies of the binding of oligosaccharides to the protein structures of lectins provide fundamental knowledge of the molecular mechanisms involved in such events. Synthetic chemists have responded to the challenges of biology; great strides have been made in creating new methodology for making available oligosaccharides required for new studies.

This research is currently rapidly accelerating. It is targeted at, among other things, developing more-refined immunological methods for diagnosis; enabling the construction of safe and specific vaccines; and creating a range of pharmaceuticals for treatment of disease. Chapters from a number of prominent researchers in these fields are collected in this book to provide an overview of the area.

Dr. Heidelberger promised to speak at the opening of the symposium on which this book is based. Unfortunately, his passing away shortly before the symposium, at the age of 103, denied us the opportunity for what would have been a remarkable experience. We decided at the symposium to dedicate this volume to the memory of Michael Heidelberger.

PER J. GAREGG
Department of Organic Chemistry
Arrhenius Laboratory
University of Stockholm
S–106 91 Stockholm, Sweden

ALF A. LINDBERG
Department of Clinical Bacteriology
Karolinska Institute
Huddinge Hospital
S–141 86 Huddinge, Sweden

October 1, 1992

Chapter 1

Michael Heidelberger
April 29, 1888—June 25, 1991

Elvin A. Kabat

Department of Microbiology, College of Physicians and Surgeons, Columbia University, New York, NY 10032

Michael Heidelberger was a Leonardo da Vinci Renaissance Man (1-4). Trained in Organic Chemistry, he became the father of Quantitative Immunochemistry[1]. He brought the precise methods of analytical chemistry to the determination of antibodies, antigens and complement on a weight basis, providing the "gold meter bar" against which the more sensitive and rapid but less precise methods of modern immunology and molecular biology such as radioimmunoassay and ELISA could be standardized and compared. He designed the first refrigerated centrifuge in desperation because his technician kept getting colds from working in the cold room. Following his design, International Equipment Co. wound a brine cooling coil around a size 2 centrifuge. Soon everyone had them. He received fifty dollars for writing the manual. He never gave up working with his own hands and was in the laboratory a few weeks before his death. His incisive mind retained its sharpness during his entire life.

Apart from his science, Michael loved music and was an accomplished, professional level clarinetist. He wrote a Wedding March which was played at his two marriages. He was always playing chamber music with friends and visiting scientists. He had a keen interest in social and political issues as well. Michael was a gifted lecturer and a great teacher, especially at the laboratory bench. He was known for his generosity, kindness and fairness, especially in giving credit to younger people. He was always accessible; the door of his laboratory was open for anyone to drop in and chat or ask his advice.

---

[1]The term Immunochemistry had been introduced earlier by Svante Arrhenius in 1907 as the title to a book of lectures.

0097–6156/93/0519–0001$06.00/0
© 1993 American Chemical Society

His first two scientific papers with F.S. Metzger at Columbia appeared in 1908. He authored over 350 papers, publishing in every decade of this century but the 1990's. At the time of his death, he left an extensive manuscript "Cross reactions of 56 bacterial polysaccharides with 24 antipneumococcal, three antisalmonella, one antiklebsiella and antimycoplasma sera" (5) that he had been working on for two years and which will be submitted for publication. He and Oswald Avery isolated and identified the type-specific substances of pneumococci as polysaccharides, opening a vast new area of immunology. He founded a school of students and colleagues who explored the immunology and immunochemistry of bacterial and other polysaccharides as well as the immunochemistry of proteins, antibodies, antigens and complement. He received a host of awards, 15 honorary degrees and 46 medals including the National Medal of Science from President Lyndon B. Johnson in 1967. I was immensely proud to have shared the Louisa Gross Horwitz Prize in 1977 with Michael and with Henry Kunkel. In 1957 the Michael Heidelberger Lecture was established at Columbia. Michael attended all but the last two and often led in the questioning.

Our association was a fortunate stroke of fate. My father had been having severe financial difficulties and to try to help him my mother began selling dresses in our apartment. Mrs. Nina Heidelberger, Michael's first wife, whom we did not know, heard about this, came in and bought some dresses. My mother learned that Michael was an Associate Professor of Biochemistry at Columbia, and was doing research. She told Mrs. Heidelberger about me and about my interest in Chemistry and in research. Mrs. Heidelberger suggested that I speak to Michael. I dropped in at his laboratory in the Department of Medicine at the College of Physicians and Surgeons several mornings between Sept. and Nov. 1932, told him about some research I was doing at City College during the afternoons and about my desire for a career in Science.

He had generally only taken postdoctoral fellows, and had never had a graduate student. He had had only a high school graduate who made solutions, helped in washing glassware, cleaned the desktop, etc. but Michael said he could not offer such a position to a college graduate. I assured him that I would gladly do all of these chores, provided that I could also work to the top level of my ability, and around Nov. 1932 he said that he would take me on in his laboratory as a laboratory helper beginning in Jan. 1933 at a salary of $90 per month. When I arrived it turned out that the Presbyterian Hospital had announced a 10 or 15 percent pay cut beginning Jan. 1, 1933 and I found my first paycheck reduced by this amount. Michael told the hospital that he had hired me at 90 dollars per month beginning in January and insisted that they honor his

agreement - they did. Soon after I started working with
Michael I applied to be a graduate student in the
Department of Biochemistry at P&S and, after an interview
and quiz by the chairman, Hans T. Clarke, was admitted in
Feb. 1933. Thus began my 59 year association with Michael
as his first graduate student and in 1937 his first Ph.D.

Michael and I had similar experiences in many ways -
he described himself as obstinate and my parents felt this
way about me. Michael's family was in very modest
circumstances and he was always working in his spare time
to make extra money assisting in summer teaching and
abstracting for Chemical Abstracts. My father had lost his
business in 1927 and we were reduced to desperate
circumstances by 1931-32 literally having no food in the
house. In 1932 he was making five dollars a week to support
a family of four.   In the summer of 1931 I worked as an
usher in the old Loews New York Theater from 1pm to 3am
seven days a week for fifteen dollars; the job gave me
several hours off in the late afternoon which were
perfectly useless.   We both wanted to be chemists and
allowed nothing to stand in our way.

We differed greatly in our relation to music, and our
musical abilities, however.   In elementary school we had
an assembly each Friday morning of the five hundred or so
students. After singing one song, the principal announced,
"We will sing the song over and will the two Kabat boys (a
cousin and I) please keep quiet." I actually liked music.
My mother played the piano very well. My talents were not
as broad as Michael's.   (For additional references and
other aspects of Michael's career see 6-14).

REFERENCES

1. Phillips, Harlan B. 1969. Michael Heidelberger Vol. I
   An oral history. Four tapes covering the period from
   Michael's birth in 1888 through 1968, pp. 1-205. Vol.
   II Documents and correspondence during this period.
   National Library of Medicine. A gold mine of documented
   information. Since I started working with Michael in
   1933, I was able to refresh my memory of many events.
   The rest of Michael's papers has been given to the
   National Library of Medicine by his grandson, Philip
   Heidelberger.

2. Heidelberger, Michael. 1984. Reminiscences - A "Pure"
   Organic Chemist's Downward Path. Immunological Reviews
   81:7-19. Reprinted and extended with permission from
   Ann. Rev. Microbiol. 31:1-12, 1977.

3. Heidelberger, Michael   1984.   Reminiscences - 2. The
   Years At P. and S.   Immunological Reviews 82:7-27.
   Reprinted with permission from Ann. Rev. Biochem. 1979,
   1-21.

4. Heidelberger, Michael   1985.   Reminiscences - 3.
   Retirement. Immunological Reviews 83:5-22. Reprinted
   and amplified with permission from The University of
   Chicago, 1981.

5. Heidelberger, M. 1991.   Cross-reactions of various
   bacteria in antipneumococcal and other antisera.
   Unfinished   manuscript   containing   much   useful
   information. Will be in Michael Heidelberger Collection
   at the National Library of Medicine and will be
   prepared for publication.

6. Kabat, E.A. 1988. Michael Heidelberger--Active at 100.
   FASEB J. 2:2233.

7. Kabat, E.A. 1981. Life in the laboratory.   Trends in
   Biochemical Sciences 6:5. Willstätter and Emil Fischer
   as seen by Hermann O.L. Fischer, Emil Fischer's son.

8. Kabat, E.A. 1975. Michael Heidelberger as   a
   Carbohydrate Chemist. Carbohydrate Research 40:1.

9. Kabat, E.A. 1978.   Life in the laboratory. Trends in
   Biochemical Sciences 3:87.

10. Lüderitz, O., Westphal, O. and Staub, A.-M.   1966.
    Immunochemistry of O and R Antigens of Salmonella and
    Related Enterobacteriaceae. Bact. Revs. 30:192-255.

11. Staub, A.-M. 1991. Letter of July 24 to Dr. Stratis
    Avrameas in response to my request to him to send me
    information about Michael's activities at the Pasteur
    Institute.

12. Wu, H., Sah, P.P.T. and Li, C.P. 1928-29. Composition
    of Antigen-Precipitin Precipitin.   Proc. Soc. Exp.
    Biol. Med. 26:737-738.

13. Kabat, E.A. 1985. Obituary: Manfred Martin Mayer
    June 15, 1916 -Sept. 18, 1984. J. Immunol. 134:655.

14. Bendiner, E.L. 1983.   Heidelberger at 95, still
    "pigheadedly productive". Hospital Practice 18:214-
    215, 220-222, 226-227, 230-234.

RECEIVED April 9, 1992

Chapter 2

How Proteins Recognize and Bind Oligosaccharides

R. U. Lemieux

Department of Chemistry, University of Alberta, Edmonton,
Alberta T6G 2G2, Canada

The results of both theoretical (Monte Carlo simulation of the hydration of
the reacting species) and experimental (probing of the combining sites
with structural congeners) studies indicate that the main component of the
driving force for the association of oligosaccharides and antibodies or
lectins can be the release to bulk of water molecules at or near the
interacting stereoelectronically complementary surfaces. All of these
binding reactions appear to involve shallow polyamphiphilic clefts at the
surface of the protein.

It has long been appreciated that water molecules contribute to the energetics of
biological associations. That they may be involved in providing the required
complementarity to the interacting surfaces is well documented (1,2). Also, the polar
groups at these surfaces and their near environments are surely strongly hydrated and
the charge delocalization can be expected to extend to several other water molecules to
build clusters specific to the molecular structures involved;

$$Ligand\ (H_2O)_x + Protein\ (H_2O)_y$$
$$K_{Assoc.}\downarrow\uparrow$$
$$(Ligand\bullet Protein)\ (H_2O)_z + nH_2O,$$

wherein $x + y = z + n$. That is, the number and arrangements of these water molecules
should vary with structural changes in the ligand (oligosaccharide) and acceptor protein
(antibody, enzyme or lectin). Since the strengths of hydrogen bonds can be as high as
6 kcal/mole, variations in the energy contents of these clusters will surely contribute to
changes in the driving force for association. It is noteworthy in this regard that water
molecules directly hydrogen-bonded to polar groups at the surface of proteins have been
detected by single crystal X-ray diffraction analysis. Delbaere and Brayer (3)
accomplished this for the complex between chymostatin and *Streptomyces griseus*
protease A. More recently, Delbaere *et al.* (private communication) demonstrated the

0097–6156/93/0519–0005$06.00/0
© 1993 American Chemical Society

presence of water molecules at the surface of crystals of the dimeric lectin IV of *Griffonia simplicifolia* (*4*) which has 243 amino acid residues in each of the two subunits (*5*). The locations of the water molecules at the combining site are displayed in Figure 1. Seven water molecules form a cluster over the shallow combining site which has an average depth of only 2.1 Å (*6*). Evidently, these water molecules form particularly strong hydrogen bonds below the periphery of the combining site with the key polar groups Ser 49, Asp 89 and Asn 135 (*7*).

The affinity of the GS-IV lectin is for the tetrasaccharide portion of the methyl glycoside, α-L-Fuc-(1d→2b)-β-D-Gal-(1b→3a)-[α-L-Fuc-(1c→4a)-β-D-GlcNAc-OMe (Le^b-OMe) (see Figure 2), which is known as the Lewis b human blood group determinant.

The Le^b-OMe•GS-IV complex at 2.15 Å resolution is displayed in Figure 3. It was truly remarkable to find that five of the ten hydroxyl groups of Le^b-OMe remain in the aqueous phase. As seen in Figure 2, these hydroxyl groups are at the 6-positions of the β-D-GlcNAc and β-D-Gal units, the 2-position of the α-L-Fuc c-residue and the 3 and 4 positions of the α-L-Fuc d-unit. That these hydroxyl groups in fact remain hydrogen-bonded to water molecules was unequivocally established by demonstrating that the lectin formed readily detected complexes with each of the mono-*O*-methyl derivatives at these positions (*8*). It is noteworthy that the results of simple hard-sphere calculations to estimate whether or not the introduction of the methyl group at these positions would destabilize the complex proved in accord with this finding. Indeed, the 2b,2c, 3d, 4d-tetra-*O*-methyl-Le^b-OMe provided a complex that is 0.5 kcal/mole more stable than the complex formed with Le^b-OMe (*8*).

The three key hydroxyl groups at positions 3b, 4b and 4c of Le^b-OMe become engaged in hydrogen bonds with polar groups at the base of the combining site (see Figure 2) which, prior to complex formation, are strongly hydrated (see Figure 1). Thus, a low-energy mechanism is provided for the displacement of these water molecules from the combining site (*9*).

It is often considered that the main driving force for complex formation is provided by the hydrogen bonds between the ligand and the protein. On the other hand, the formation of these bonds may be mainly to meet the stereoelectronic complementarity necessary for the complex to form. Indeed, this investigation was instigated by the consideration that the main source of the driving force may reside in the changes in energy that result from changes in interactions between water molecules that hydrate the reacting species (*10*). In order to test this idea, an effort was made to simulate the hydration of methyl α-L-fucopyranoside (*11*). As was expected, the theoretical model indicated that the associations with the water molecules were at much shorter range with the polar groups of the simple glycoside than with its nonpolar regions. This notion became strongly reinforced on finding that the thermodynamic parameters for the binding of Le^b-OMe by GS-IV (Figure 4) could be importantly affected by the deoxygenation of hydroxyl groups that, in the complex (see Figure 2), are near the periphery of the combining site but hydrogen-bonded to water molecules. This observation has the important consequence of proving that efforts to establish the size of combining sites by way of probing experiments with congeners of the epitope may lead to the conclusion that the size is much larger than it actually is.

The differential changes in free energy for all the monodeoxy derivatives of Le^b-OMe (*12*) are reported in Figure 4. Of particular importance is the finding that deoxygenation of a hydroxyl groups that do not become directly hydrogen-bonded to the lectin (for example, positions 2c and 3d) can have as strong an effect on the stability of the complex as does the deoxygenated hydroxyl groups that do become hydrogen-bonded to amino acid residues of the combining site (for example, those at positions 3c and 2d, see Figure 2).

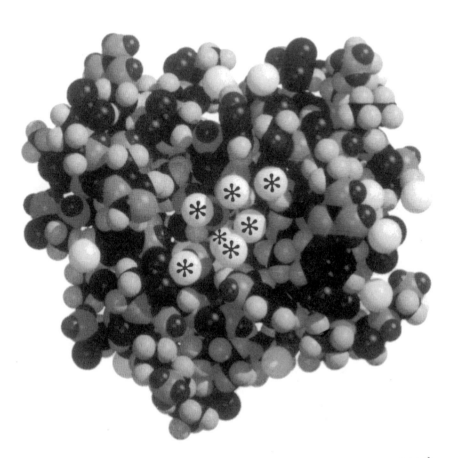

Figure 1. Fragment of the native GS-IV lectin's X-ray crystal structure at 2.15 Å resolution to display the 62 amino acid residues at and about the combining site for Le^b-OMe. The asterisked, light-colored spheres are the oxygen atoms of water molecules hydrogen bonded to the key polar groups of the combining site.

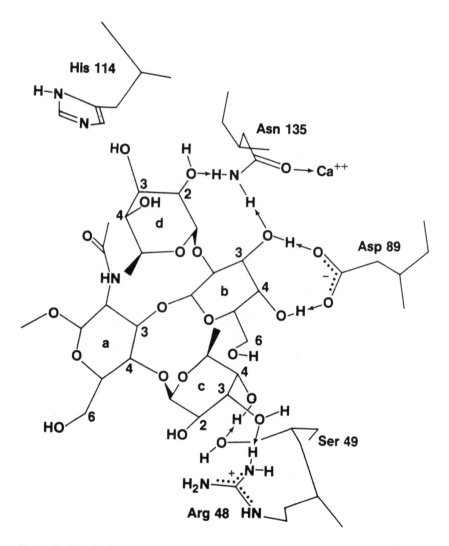

Figure 2. The hydrogen bonded network in the complex formed by the Le^b-OMe tetrasaccharide and the combining site of the GS-IV lectin (8). Note that three of the hydrogen bonds are deep in the shallow combining site and sheltered from competition with water molecules and that two other hydrogen bonds are at the periphery of the combining site and exposed to water.

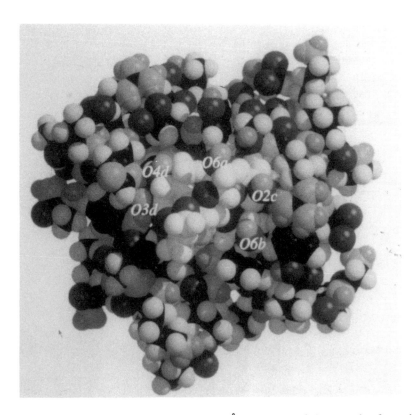

Figure 3. The X-ray crystal structure at 2.15Å resolution of the complex formed by Le^b-OMe with the GS-IV lectin. The darker atoms represent 62 of the 243 amino acid residues present in each subunit of the dimeric lectin. The labeled atoms represent the five hydroxyl groups of Le^b-OMe that remain in contact with the aqueous phase.

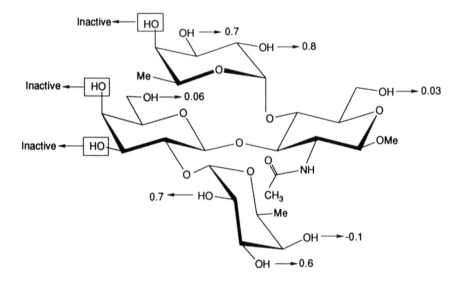

$$^*\Delta\Delta G^\circ = \Delta G^\circ_{\text{Monodeoxy-Le}^b\text{-OMe}} - \Delta G^\circ_{\text{Le}^b\text{-OMe}}$$

Figure 4. Differential changes in free energy (relative to Le^b-OMe in kcal/mole at 298 °C that occur as the result of replacing a hydroxyl group by a hydrogen atom on binding by the GS-IV lectin (12). It is to be noted with reference to Figure 2 that the change in free energy is at near zero when the substitution is at the position most remote from the protein (i.e., OH-6a). In this case, the differential change in enthalpy was also near zero.

As previously reported (*12*), deoxygenations of Le[b]-OMe produce decreases in enthalpy which are much greater than the changes in free energy. It had long been recognized that, as reviewed by Lumry and Rajender (*13*), such enthalpy-entropy compensation phenomena most likely result from a unique property of water (*13*). It could be expected, therefore, that the compensations found for the binding of Le[b]-OMe and its deoxy congeners by the GS-IV lectin occur because changes in the structure of the ligand cause changes in the structures (of course time averaged) of the hydration shells. It was proposed, for reasons to be detailed below, that the associations lead to a decrease in enthalpy, because the hydrogen bonds between water molecules near the surfaces of the reacting molecules are, on the average, weaker than the hydrogen bonds of bulk water. On the other hand, the surface water molecules are expected to be less ordered than those in bulk solutions and, on being released to bulk, produce the compensating decrease in entropy.

The reason why the water molecules at the interacting surfaces are believed to be less strongly hydrogen-bonded and in greater disorder than in bulk was attributed to the hydration of highly dispersed regions of amphiphilicity (*12*). Such surfaces, for which we have proposed the term "polyamphiphilic" (*14*) are present in virtually all organic molecules and the particular array in a given molecule is fully characteristic of that molecule. The spacings between the polar groups do not necessarily coincide well with the distances required for the formation of hydrogen bonds between the water molecules in the inner hydration shells that cover the polyamphiphilic surfaces. Also, hydrogen bonds are highly directional in character and, in order to achieve maximum strength, may tend to direct the water molecules away from an adjacent nonpolar group. The compromise, under thermal agitation, must result in hydrogen bonds that are weaker on the average than those between water molecules in bulk solution. For these reasons, it was surmised (*9,10*) that the density of water over the nonpolar regions of a polyamphiphilic surface must be lower than that over the polar groups.

In order to achieve a better appreciation of how water molecules organize themselves over a polyamphiphilic surface, it was decided to examine Monte Carlo simulations of the hydration at 300° of the supremely polyamphiphilic surfaces of the Le[b]-OMe tetrasaccharide and the combining site of the GS-IV lectin (Figure 5). It was anticipated that thereby an insight may be gained on how the release of water molecules from the interacting surfaces affects their association to form the Le[b]-OMe•GS-IV complex. In order to reduce the supercomputer CPU time requirements, a number of arbitrary simplifications were made which are not expected to adversely affect the conclusions reached (*6,10*).

Our first effort was to examine the hydration of Le[b]-OMe (*10*). As expected, it became clearly apparent that in aqueous solution the nonpolar C–H groups are at least one Å removed from the nearest water molecule. We have now studied the hydration of the combining site of the GS-IV lectin (*6*).

The simulation of the hydration of the protein surface displayed in Figure 5 used 250 water molecules at unit density and 300°. As for Le[b]-OMe (*10*), the energy minimization procedures in the Monte Carlo calculations were based on Clementi's potentials but using a Cray X-MP computer and additional class assignments for atoms and groups present in the protein structure (Figure 5). A special procedure was used for implementing boundary conditions such that the surface water molecules of the water cluster appear to be interacting with water and not facing void space (*6*). Equilibrium was reached after 1000 Monte Carlo cycles in which every molecule was tentatively moved once, as previously described (*10*). Thereafter, samples were taken after every 500 cycles up to a total of 5000 cycles in order to obtain the data presented in Figures 6-8.

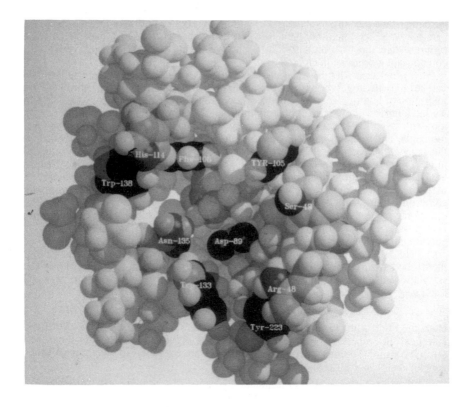

Figure 5. The combining site of the GS-IV lectin without the water of hydration to depict its polyamphiphilic character. Note the abundance of aromatic residues about its periphery.

The structure of the GS-IV•$(H_2O)_{250}$ complex after 5000 sampling cycles is presented in Figure 6 in order to provide a glimpse on how the water molecules become organized over the polyamphiphilic protein structure. It is at once seen that at that particular "moment" the water structure near the protein appears "porous" and highly disordered whereas that toward the top of the cluster already appears more dense and orderly.

The effectively "porous" quality of the layer of water over the protein surface can be better appreciated from Figures 7 and 8. These data clearly indicate that the water molecules which become bound to the polar groups of the protein tend to accept further water molecules in directions that effectively leave void space over the nonpolar regions of the polyamphiphilic surface. It seems evident, therefore, that complex formation will lead to important stabilization of the system when these water molecules are released to form higher density bulk water. The nature and strength of the condensation forces are under investigation.

The above conclusion appears particularly attractive because it provides an answer to the perplexing problem of why rather small hydrophilic surfaces can associate in aqueous solution. In the case of the Le^b-OMe•GS-IV complex, the interacting complementary hydrophilic surfaces (Figures 5 and 9) have areas of only about 237 Å^2. Nevertheless, the stabilization of the system gained by complex formation is sufficiently great to lead to an essentially quantitative formation of the complex in water at micromolar concentrations. Since all biological molecules possess polyamphiphilic structural features, the proposal obviously has extremely broad and diverse significance to biological processes.

Our probing of the combining sites of a wide number of antibodies and lectins has shown their epitopes to be similar in kind to that established for Le^b-OMe which is displayed in Figure 9. Indeed, almost the same surface of Le^b-OMe is recognized by a monoclonal anti-Lewis a antibody (15). Also, a monoclonal antibody was found to recognize the H-type 2 human blood group determinant in a manner remarkably similar to that employed by the lectin I of *Ulex europaeus* (14). Of course, these are chance observations. That different regions of an oligosaccharide's surface can serve as epitope is provided by the structures displayed in Figure 10 for the Lewis a (15), Lewis b (16) and B human (17) blood group antigenic determinants that are recognized by monoclonal antibodies. It is seen that two monoclonal anti-Lewis a antibodies recognize very different features of the Lewis a trisaccharide. Also, the surface of Le^b-OMe bound by the monoclonal anti-Lewis b antibody is very different to that recognized by the GS-IV lectin (Figure 9). Most important to note is that all four epitopes have polyamphiphilic features and possess polar groups that can be expected to establish key polar interactions that are sheltered from hydrolytic action once the complex has formed.

To summarize, we contend that the release to bulk of those water molecules that solvate complementary polyamphiphilic surfaces makes an important contribution to their association in aqueous media. The formation of a complex inevitably faces an important entropy barrier for the organization and immobilization of the reacting species. Also, as proposed above, the reduction in area of polyamphiphilic surfaces facing water can be expected to reduce the number of water molecules that are less ordered than the molecules of bulk water. However, these entropy barriers should be importantly compensated by the release to bulk of the water molecules that are strongly hydrogen-bonded to the reacting surfaces as was seen for the combining site of the GS-IV lectin in Figure 1. Furthermore, contributions akin to hydrophobic effects may occur especially in view of the high preponderance of the large aromatic residues that are often found about the peripheries of the combining sites. This feature of the combining site of the GS-IV lectin is displayed in Figure 5. Similarly high abundances

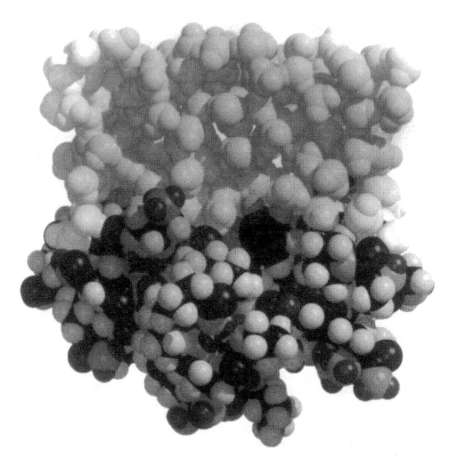

Figure 6. The cluster of 250 water molecules placed uniformly over the combining site of the GS-IV lectin (area $\simeq$ 237 $Å^2$) with unit density at 300 $°$ C and after 5000 Monte Carlo sampling cycles. The dimensions of the cluster are near 29.6 $\times$ 21.3 Å along the x and y axes, respectively, and 25.7 Å high.

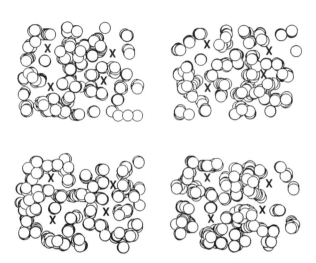

Figure 7. Looking down toward the receptor site of the GS-IV lectin. Circles represent the oxygen atoms of the water molecules that were found within 2.5 Å of the lectin after each of ten stages, each taken after 500 Monte Carlo sampling cycles, in four different simulations. The regions marked by X that appear free of water molecules are over nonpolar regions near the periphery of the combining site.

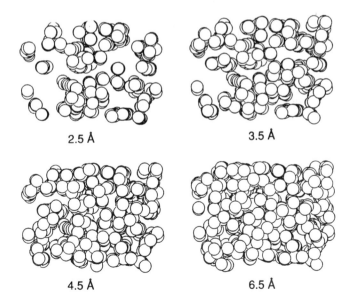

2.5 Å   3.5 Å

4.5 Å   6.5 Å

Figure 8. As for Figure 7 except that the energetically preferred positions of the water molecules are displayed at increasing distances from the surface of the lectin.

Figure 9. The hydrophilic polyamphiphilic surface of the Le^b-OMe that is stereoelectronically complementary to the combining site of GS-IV. The hydroxyl groups at positions 3b, 4b, and 4c become hydrogen-bonded as proton donors deep within the combining site. The hydroxyl groups at positions 3c and 2d become hydrogen-bonded as proton acceptors near the periphery of the combining site (see Figure 2).

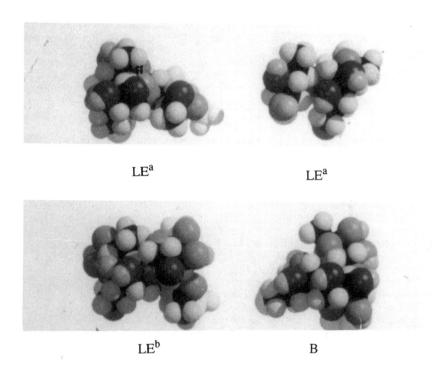

LE^a   LE^a

LE^b   B

Figure 10. The hydrophilic polyamphiphilic antigenic groups for two anti-Lewis a, an anti-Lewis b, and an anti-B monoclonal antibodies. Note that the surfaces recognized are different for the two anti-Lewis a antibodies and that recognized by the anti-Lewis b antibody is very different to that bound by the GS-IV lectin (Figure 9).

of aromatic groups have been discovered for the combining sites of an influenza virus haemagglutinin (*18*) and an antibody specific for a trisaccharide epitope of pathogenic *Salmonella* (*2*). A recent analysis of the combining sites of three antilysozyme Fabs revealed the occurrence of a large number of aromatic residues, particularly that of tyrosine (*19*).

In conclusion, I wish to acknowledge with thanks the contributions to the research that I have described by my colleagues. The computer programs for the Monte Carlo simulations were written and executed by Dr. Helmut Beierbeck who also prepared all the computer graphics for the various Figures. Professor Louis Delbaere generously allowed the use of the crystallographic coordinates at 2.15 Å resolution prior to publication. The thermodynamic studies were performed by Mrs. Mimi Bach under the supervision of Dr. Ulrike Spohr.

The research was supported by my operational grant (OGP 172) from the Natural Sciences and Engineering Research Council of Canada. The National Research Council of Canada has granted permission to reproduce all Figures used in this paper which have or may be published in the Canadian Journal of Chemistry.

Literature Cited

1. Quiocho, F.A.; Wilson, D.K.; Vyas, N.K. *Nature*, **1989**, *340*, 404-407.
2. Cygler, M.; Rose, D.R.; Bundle, D.R. *Science*, **1991**, *253*, 442-445.
3. Delbaere, L.T.J.; Brayer, G.D. *J. Mol. Biol.* **1985**, *183*, 89-103.
4. Shibata, S.; Goldstein, I.J.; Baker, D.A. *J. Biol. Chem.* **1982**, *257*, 9324-9329.
5. Nikrad, P.V.; Pearlstone, J.R.; Carpenter, M.R.; Lemieux, R.U.; Smillie, L.B. *Biochem. J.* **1990**, *272*, 343-350.
6. Beierbeck, H.; Vandonselaar, M.; Delbaere, L.T.J.; Lemieux, R.U. *Can. J. Chem.*, to be submitted.
7. Delbaere, L.T.J.; Vandonselaar, M.; Prasad, L.; Quail, J.W.; Nikrad, P.V.; Pearlstone, J.R.; Carpenter, M.R.; Smillie, L.B.; Spohr, U.; Lemieux, R.U. *Can. J. Chem.* **1990**, *68*, 1116-1121.
8. Nikrad, P.V.; Beierbeck, H.; Lemieux, R.U. *Can. J. Chem.* **1992**, *70*, 241-253.
9. Lemieux, R.U. *Chem. Soc. Rev.* **1989**, *18*, 347-374.
10. Beierbeck, H.; Lemieux, R.U. *Can. J. Chem.* **1990**, *68*, 820-827.
11. Lemieux, R.U. *Proceedings of the VIIIth International Symposium on Medicinal Chemistry*, Swedish Pharmaceutical Press, Stockholm, **1985**, *1*, 329-351.
12. Lemieux, R.U.; Delbaere, L.T.J.; Beierbeck, H.; Spohr, U. *Ciba Foundation Symposium No. 158*, Wiley, Chichester, London, U.K. **1991**, 231-248.
13. Lumry, R.; Rajender, S. *Biopolymers*, **1970**, *9*, 1125-1227.
14. Spohr, U.; Paszkiewicz-Hnatiw, E.; Morishima, N.; Lemieux, R.U. *Can. J. Chem.* **1992**, *70*, 254-271.
15. Lemieux, R.U.; Hindsgaul, O.; Bird, P.; Narasimhan, S.; Young, Jr., W.W. *Carbohydr. Res.* **1988**, *178*, 293-305.
16. Spohr, U.; Hindsgaul, O.; Lemieux, R.U. *Can. J. Chem.* **1985**, *63*, 2644-2652.
17. Lemieux, R.U.; Venot, A.P.; Spohr, U.; Bird, P.; Mandal, G.; Morishima, N.; Hindsgaul, O.; Bundle, D.R. *Can. J. Chem.* **1985**, *63*, 2664-2668.
18. Weis, W.; Brown, J.H.; Cusack, S.; Paulson, J.C.; Skehel, J.J.; Wiley, D.C. *Nature*, **1988**, *333*, 426-431.
19. Padlan, E.A. *Proteins: Structure, Function and Genetics*, **1990**, *7*, 112-124.

RECEIVED March 28, 1992

Chapter 3

Synthesis and Immunological Properties of Glycopeptide T-Cell Determinants

Morten Meldal[1], Søren Mouritsen[2], and Klaus Bock[1]

[1]Department of Chemistry, Carlsberg Laboratory, Gamle Carlsberg Vej 10, 2500 Valby, Copenhagen, Denmark
[2]Institute of Experimental Immunology, University of Copenhagen, Nørre Allé 71, 2100 Copenhagen 0, Denmark

A branched pentasaccharide derived from degradation of amylopectin was transformed into a glycosylated amino acid building block, 5, suitably protected and activated for continuous flow solid phase peptide synthesis. The building block was incorporated into peptide T-cell determinants, hen egg lysozyme 81-96 and artificial MP7, to be used in immunological studies. Mice immunized with glycosylated MP7 elicit high antibody titers against the carbohydrate part of the glycopeptide.

Introduction. Antibodies directed towards carbohydrate structures can in general only be raised if the carbohydrate is suitable activated and subsequently conjugated to a protein carrier molecule as e. g. Keyhole Limpet Hemocyanin or BSA, prior to immunization.

Conjugation is required since only proteins seem to be able to initiate significant T-cell responses, while all kinds of antigen molecules can be recognized by the B lymphocyte immunoglobulin receptor. This limitation in the T-cell repertoire is probably due to the fact that T-cells recognize only antigen fragments bound to the Major Histocompatibility Complex (MHC) molecules and that carbohydrate structures alone seem unable to bind to these molecules (1). When T-cells are not involved in an immune response only weak antibody titers are produced.

With the recognition of peptide epitopes as stimulators for T-cells we have envisaged that glycopeptides with an oligosaccharide suitable linked to a known T-cell epitope could form the basis for production of antibodies against the carbohydrate structure. The preparation of glycopeptides is a prerequisite for testing this hypothesis and will be reported in the following.

In studies of the structure and function of glycoconjugates, easy solid phase synthesis of a variety of glycopeptides is indispensable. A building block strategy in which glycosylated amino acids are incorporated in a sequential solid phase synthesis has been found to be most efficient (2-16). The previously described methods for solid phase synthesis all used more or less permanent orthogonal protection of the α-amino group, the glycan part and the α-carboxyl group. After glycosylation the α-carboxyl group was deprotected, purified and subsequently activated for solid phase synthesis. Mostly Fmoc-amino acids protected at the α-carboxyl group as t-butyl (4), phenacyl (5) or allyl esters (17) have been employed. Valuable glycosylated amino acid derivative is, however, lost during these operations and a tendency to use in situ coupling which does not allow characterization of the coupling reagent has prevailed. A few reports describe the glycosylation of weakly activated amino acid derivatives for solution phase synthesis (18, 19). Solid phase synthesis requires, however, high activation of the coupling reagent. We have recently demonstrated the utility of the pentafluorophenyl (Pfp) ester for the general temporary α-carboxyl protection during glycosylation in both O- (9, 13) and N-glycopeptide synthesis (11). The versatility of the method allowed the multiple column peptide synthesis of a large number of mucin type glycopeptides (15). The Pfp esters are simultaneously highly activated for aminolysis yet sufficiently stable to survive under glycosylation and purification conditions.

In this work we describe the versatility of the method with further examples including O-glycosylation of serine and N-glycosylation of asparagine in the synthesis of glycosylated peptides expected to be active as T-cell determinants in immunological studies.

Peptide synthesis was carried out on a custom made continuous flow peptide synthesizer. Active esters were delivered from an in-line solid sampler and a solid phase spectrophotometer was used to monitoring the coloration of the resin with Fmoc-amino acid-Dhbt esters (20). A catalytic amount of Dhbt-OH was added when Pfp esters were used. Fmoc-cleavages were performed by treatment with 50% morpholine in DMF in the case of aliphatic O-glycopeptide synthesis or with 20% piperidine for appropriate periods of time (30-70 min) for N-glycopeptides. It should be noted that morpholine was found to be insufficiently basic in the Fmoc-cleavage reactions of longer peptides and even with prolonged reaction times some incomplete deprotection was observed. Kieselguhr supported polydimethyl acrylamide resin was used as the solid phase and DMF as the solvent.

The **methodology** we have employed for preparation of O-glycopeptides is based on the glycosylation of easily available Fmoc-Ser-O-Pfp (*21, 22*), **1**, which can be synthesized by carbodiimide activation with either diisopropyl carbodiimide or dicyclohexyl carbodiimide in THF. Similar preparations can be performed with Fmoc-Thr-OH and Fmoc-Tyr-OH (*23*). The yield after chromatographic purification varies from 64-75% and compound **1** may be isolated by direct crystallization. It must, however, be emphasized that residual DCU can not be tolerated in the glycosylation reactions. We have therefore developed purification procedures for Pfp esters on reversed phase HPLC (*23*), dried silica normal phase MPLC, VLC or flash chromatography (*9, 13*).

O-Glycopeptide synthesis: From degradation studies on starch with the enzyme Termamyl it was possible to isolate a branched pentasaccharide **2** (*24*). This was conjugated into a T-cell determinant from hen egg lysozyme and the artificial determinant, MP7 (*25*), in an attempt to raise antibodies towards the saccharide and to study the binding of glycosylated T-cell determinants to the MHC class II molecules. The pentasaccharide was peracetylated (**3**) and transformed by reaction with titanium tetrabromide into the glycosyl bromide, **4**, in 79% overall yield. Compound **4** was subsequently used to glycosylate **1** using silver trifluoromethanesulfonate as a promoter yielding 42% after silica gel chromatography of the glycosylated building block, **5**. Compound **5** was used in a synthesis of the glycosylated peptide fragment 81-96 from hen egg lysozyme known to be a T-cell determinant. The structures of the glycopeptide products **7** and **8** were confirmed by amino acid sequencing and 1- and 2D-NMR spectroscopy (Tables 1 and 2). The size of all $J_{H\alpha\text{-}NH\alpha}$ indicated a random and non α-helical structure of the peptide backbone (*26*). Compound **5** was furthermore used in the synthesis of the artificial and immunologically characterized amphiphatic α-helical T-cell determinant, MP7 (**9**) (*25*), yielding the terminally glycosylated product, **10**. The fully assigned ^{1}H-NMR spectra in Table 1 clearly indicate for compound **10** a large content of α-Helical conformation of the peptides with $J_{H\alpha\text{-}NH\alpha}$ ~ 5 Hz and no observation of Noe correlations between consecutive backbone amide protons (*27*).

N-Glycopeptides can be synthesized from the Fmoc-Asp(Cl)-O-Pfp, **11**, which can be obtained quantitatively in two steps from Fmoc-Asp(tBu)-O-Pfp (*11*). Thus glycosylamine **12**, synthesized from GlcNAc by successive reaction with ammonium carbonate, Fmoc-O-Su and acetic anhydride followed by cleavage of the Fmoc group by piperidine in THF, was reacted selectively with the acid chloride **11**. This afforded the fully protected

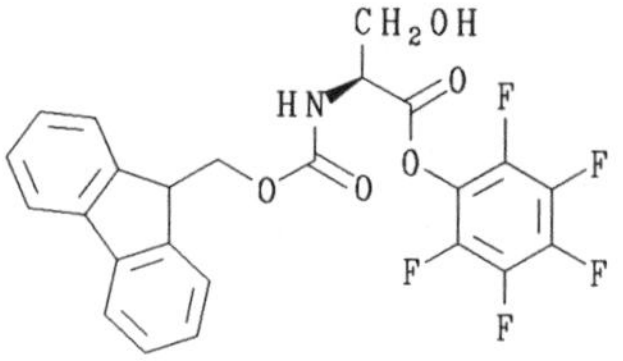

1

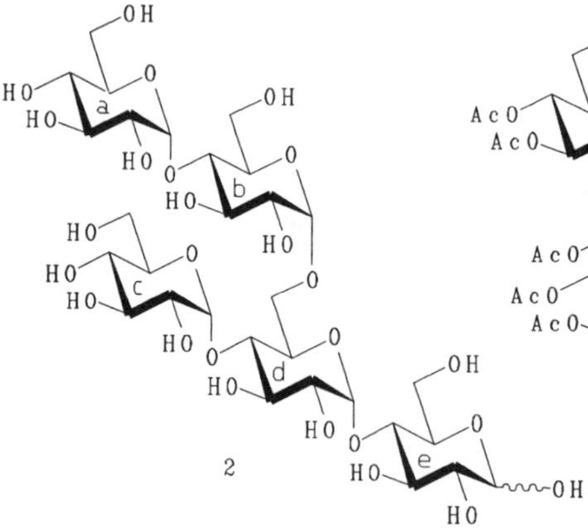

2

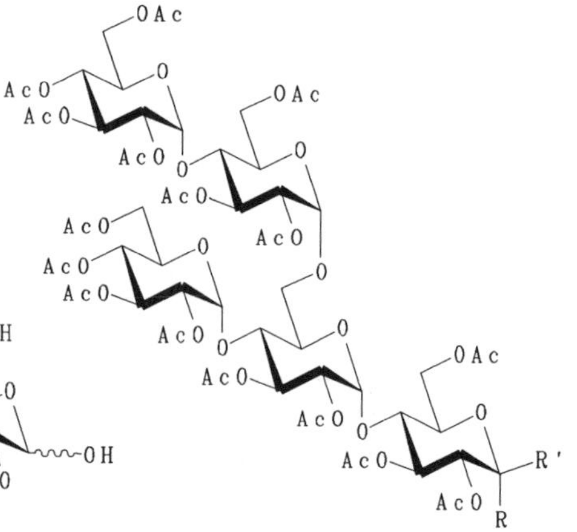

3 : R , R '= H , O Ac
4 : R = B r , R '= H

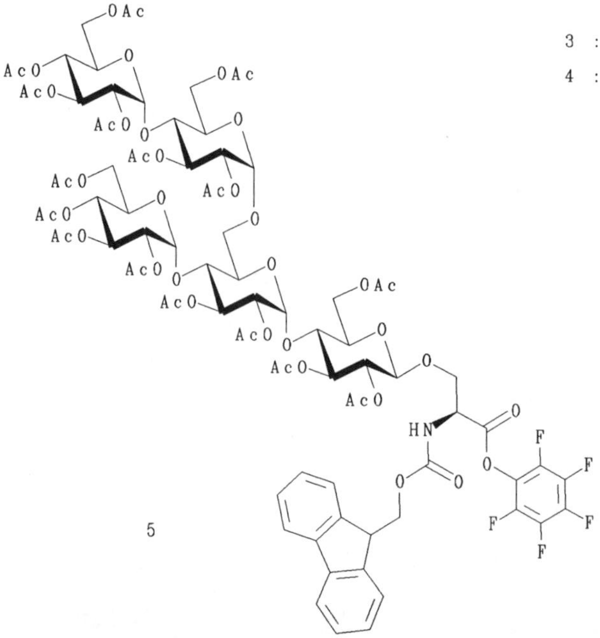

5

$$\overset{\displaystyle R^1}{|}$$
H—Ser—Ala—Leu—Leu—Ser—Ser—Asp—Ile—Thr—

$$\overset{\displaystyle R^2}{|} \qquad \overset{\displaystyle R^3}{|}$$
Ala—Ser—Val—Asn—Cys—Ala—Lys—Tyr—CO_2H

6   : $R^1 = H$, $R^2 = H$, $R^3 = H$

7   : $R^1 = H$, $R^2 = Glc_5$, $R^3 = H$

8   : $R^1 = Glc_5$, $R^2 = H$, $R^3 = H$

14   : $R^1 = H$, $R^2 = H$, $R^3 = GlcNAc$

$$\overset{\displaystyle R^1}{|}$$
H—Ser—Pro—Glu—Leu—Phe—Glu—Ala—Leu—

Gln—Lys—Leu—Phe—Lys—His—Ala—Tyr—CO_2H

9   : $R^1 = H$

10   : $R^1 = Glc_5$

Table 1. Selected ^{1}H-NMR data (p.p.m. values and coupling constants in the parameters for the saccharide parts of the synthesized compounds. The letters denoting the respective units in oligosaccharides are indicated in scheme 1. Compound **2** has been characterized as the β-methyl glycopyranoside

	H1 ($J_{1,2}$)	H2 ($J_{2,3}$)	H3 ($J_{3,4}$)	H4 ($J_{4,5}$)	H5 ($J_{5,6}$)	H6 ($J_{5,6'}$)	H6' ($J_{6,6'}$)
2a	5.40 (3.6)	3.56 (9.8)	3.68 (9.8)	3.38 (9.8)	3.71	3.77	3.84
2b	4.96 (3.6)	3.58 (9.8)	3.98 (9.8)	3.58 (9.8)	3.84	3.80	3.84
2c	5.39 (3.6)	3.57 (9.8)	3.68 (9.8)	3.40 (9.8)	3.71	3.77	3.84
2d	5.36 (3.8)	3.62 (9.8)	3.92 (9.8)	3.64 (9.8)	4.00	3.84	3.94
2e	4.38 (8.4)	3.28 (9.8)	3.75 (9.8)	3.62 (9.8)	3.58	3.80	3.96
4a	5.43 (3.5)	4.92	5.36	5.12	4.06	4.29	4.60
4b	5.32 (3.8)	4.80 (10.0)	5.49	4.03	4.15	4.29	4.12
4c	5.43 (3.6)	4.90	5.40	5.08	4.07	4.45	4.54
4d	5.25 (4.0)	4.68 (10.5)	5.45	3.96	4.08	3.98	3.89
4e	6.56 (4.0)	4.77 (9.5)	5.67	4.01	4.34	4.45	4.28
5a	5.43 (3.6)	4.90 (9.0)	5.37 (9.5)	5.10 (9.8)	4.03	4.56 (2.4)	4.28 (12.2)
5b	5.40 (3.8)	4.79 (10.0)	5.48 (9.4)	4.03	4.03	4.29	4.09
5c	5.43 (3.6)	4.91 (9.0)	5.41 (9.0)	5.10 (9.8)	4.07	4.50 (2.6)	4.33 (12.6)
5d	5.25 (4.0)	4.68 (9.5)	5.47 (9.4)	4.01	4.03	4.40 (3.8)	4.05 (11.0)
5e	4.53 (8.4)	4.83 (8.8)	5.31 (9.0)	3.94 (9.0)	3.64 (1)	4.19 (4)	4.48 (8)
7a	5.33 (3.5)	3.57	3.71 (9.9)	3.39 (9.9)	3.71	3.80	3.90
7b	4.96 (3.5)	3.57	3.98	3.64	3.84	3.81	3.96
7c	5.33 (3.5)	3.57	3.71 (9.9)	3.41 (9.9)	3.71	3.80	3.90
7d	5.28 (3.8)	3.62	3.93	3.67	3.99	3.86	3.96
7e	4.47 (8.4)	3.32	3.77	3.56	3.58	3.78	3.98
8a	5.35 (3.4)	3.58	3.70	3.42	3.71	3.84	3.91
8b	4.96 (3.8)	3.57	3.98	3.59	3.85	3.86	3.96
8c	5.35 (3.4)	3.60	3.68	3.44	3.73	3.84	3.92
8d	5.32 (3.3)	3.60	3.95	3.62	4.01	3.88	3.96
8e	4.50 (8.0)	3.35	3.77	3.62	3.60	3.87	3.98
14	5.02 (9.5)	3.79 (9.5)	3.59 (9.0)	3.46	3.45	3.82	3.73
10a	5.33 (3.8)	3.57	3.68	3.41	3.71	3.82	3.90
10b	4.95 (3.6)	3.52	3.98	3.57	3.80	3.83	3.97
10c	5.33 (3.8)	3.58	3.68	3.41	3.74	3.80	3.90
10d	5.30 (3.5)	3.67	3.96	3.63	3.99	3.82	3.94
10e	4.48 (8.4)	3.34 (10.0)	3.78	3.62	3.58	3.87	3.96

SOURCE: Data are from reference 24.

Table 2. Selected ^{1}H-NMR data for the peptide parts of the synthesized compounds. [a] Compound **9** was dissolved in 30% CD_3CO_2D in H_2O due to low water solubility

	6	**7**	**8**	**14**	**9**[a]	**10**
Serα	4.140 (4.5)	4.130	4.135	4.130	Serα	4.290
β	4.002 (14.0)	3.920	4.240	4.005	β	4.060
β'	3.962 (6.0)	3.920	4.240	4.970	β'	4.190
AlaN^{α}H	8.282 (5.7)	8.240 (6.0)	8.220	8.240 (6.0)	Proα	4.503
α	4.255	4.255 (6.8)	4.255	4.260	β	2.034	2.325
β	1.320 (7.0)	1.320	1.320 (7.0)	1.340 (7.0)	β'	2.381	1.870

Table 2. Continued

	6	7	8	14		9[a]	10
LeuN$^\alpha$H	8.255 (6.8)	8.240 (5.5)	8.240 (5.5)	8.348 (6.0)	γ	1.990	2.010
α	4.415	4.395	4.420	4.410	γ'	1.990	2.025
β	1.630	1.630	1.630	1.620	δ	3.545	3.666 (14.9)
γ	1.620	1.610	1.610	1.620	δ	3.545	3.763 (6.7)
δ	0.918	0.920	0.920	0.920	GluN$^\alpha$H	8.003 (4.9)	8.110 (5.0)
LeuN$^\alpha$H	8.270 (6.0)	8.260 (6.3)	8.240 (5.5)	8.280 (6.0)	α	4.122	4.185
α	4.330	4.305	4.320	4.320	β	2.054	2.060
β	1.595	1.590	1.595	1.600	β'	2.054	1.970
γ	1.620	1.610	1.620	1.620	γ	2.30	2.320
δ	0.870	0.870	0.870	0.870	LeuN$^\alpha$H	7.961 (6.3)	8.245 (6.5)
SerN$^\alpha$H	8.278 (6.5)	8.278 (6.3)	8.290 (-)	8.285 (6.5)	α	4.304	4.275
α	4.424	4.430	4.420	4.420	β	1.563	1.480
β	3.904	3.850	3.850	3.850	β'	1.563	1.480
β'	3.836	3.820	3.820	3.820	γ	1.545	1.520
SerN$^\alpha$H	8.248 (7.0)	8.270 (5.5)	8.275	8.270 (7.0)	δ	0.830	0.826 (5.0)
α	4.435	4.450	4.450	4.440	δ'	0.895	0.890
β	3.860	3.850	3.86	3.815	PheN$^\alpha$H	7.980 (7.7)	8.185 (5.0)
β'	3.843	3.880	3.88	3.845	α	4.604	4.515
AspN$^\alpha$H	8.373 (7.6)	8.315 (7.1)	8.370 (-)	8.600 (8.0)	β	2.974	3.07
α	4.590 (6.0)	4.692 (7.4)	4.720	4.750	β'	3.168	3.11
β	2.860 (16.0)	2.730 (15.5)	2.845	2.723 (16.0)	Ar	7.22	7.22 (7.0)
β'	2.750 (7.6)	2.850	2.725	2.890	Ar	7.32	7.32
IleN$^\alpha$H	8.031 (7.0)	8.020 (7.9)	8.044 (7.0)	8.030 (7.0)	GluN$^\alpha$H	8.045 (5.6)	8.163 (5.5)
α	4.210	4.210	4.210	4.210	α	4.199	4.185
β	1.910	1.900	1.920	1.910	β	2.029	2.03
γ	1.198	1.180	1.190	1.200	β^1	2.029	1.93
γ'	1.460	1.450	1.460	1.450	γ	2.35	2.38
γ"	0.905	0.895	0.905	0.910	AlaN$^\alpha$H	8.030 (5.2)	8.110 (5.0)
δ	0.860	0.860	0.860	0.860	α	4.189	4.167
ThrN$^\alpha$H	8.126 (7.2)	8.110 (7.6)	8.126 (7.5)	8.135 (7.0)	β	1.425	1.381 (7.0)
α	4.305	4.330	4.310	4.310	LeuN$^\alpha$H	8.302 (5.6)	8.022 (5.5)
β	4.205	4.210	4.210	4.200	α	4.279	4.209
γ	1.196	1.195 (6.3)	1.195 (6.3)	1.200 (6.3)	β	1.746	1.54
AlaN$^\alpha$H	8.660 (5.5)	8.660 (6.0)	8.640 (5.0)	8.660 (5.5)	β'	1.746	1.65
α	4.378	4.380	4.380	4.380	γ	1.620	1.630
β	1.375 (7.0)	1.375 (6.8)	1.370 (7.0)	1.380 (7.0)	δ	0.860	0.859 (5.5)
SerN$^\alpha$H	8.252 (6.5)	8.230 (5.5)	8.230	8.228 (5.8)	δ'	0.920	0.916
α	4.440	4.465	4.460	4.450	GluN$^\alpha$H	8.623 (5.9)	8.476 (5.5)
β	3.854	3.880	3.88	3.820	α	4.442	4.270
β'	3.914	3.920	3.92	3.820	β	2.054	1.98
ValN$^\alpha$H	8.078 (7.5)	7.995 (7.9)	8.069	8.060 (7.2)	β'	1.946	1.93
α	4.128 (5.5)	4.130	4.140	4.130	γ	2.30	2.30
β	2.100 (6.0)	2.075	2.100	2.100	NH	7.330	7.41
γ	0.920	0.915	0.930	0.915	NH	6.844	6.82

Continued on next page

Table 2. Continued

	6	7	8	14		9[a]	10
AsnN$^\alpha$H	8.448 (7.6)	8.440 (6.3)	8.430 (7.0)	8.363 (7.0)	LysN$^\alpha$H	7.941 (5.9)	8.053 (5.0)
α	4.460	4.605	4.770	4.750	α	4.174	4.176
β	2.830 (7.2)	2.830	2.845	2.790 (16.5)	β	1.726	1.75
β'	2.920 (16.0)	2.930	2.930	2.890 (6.0)	β'	1.726	1.75
N	7.508	7.585	7.539	8.155 (8.5)	γ	1.40	1.41
N	6.980	6.890	6.872	-	γ'	1.35	1.32
CysN$^\alpha$H	7.978 (7.9)	8.025 (7.9)	7.922	7.855 (7.9)	δ	1.676	1.63
α	4.570	4.600 (4.9)	4.600	4.510	ε	3.095 (5.9)	2.94
β	2.915 (15.0)	2.915 (13.4)	2.920	2.920 (14.0)	LeuN$^\alpha$H	7.749	7.888 (6.5)
β'	3.135 (5.2)	3.145	3.145	3.153 (5.1)	α	4.224	4.241
AlaN$^\alpha$H	8.196 (5.8)	8.242 (6.0)	8.196 (4.8)	8.192 (5.0)	β	1.552	1.42
α	4.340	4.350 (6.8)	4.360	4.350	β'	1.552	1.52
β	1.396 (7.0)	1.400	1.400 (7.0)	1.410 (7.0)	γ	1.535	1.496
LysN$^\alpha$H	8.073 (7.5)	8.020 (7.9)	8.045 (7.5)	8.030 (6.5)	δ	0.775	0.785 (6.0)
α	4.230	4.200	4.200	4.215	δ'	0.850	0.862 (6.0)
β	1.635	1.650	1.635	1.920	PheN$^\alpha$H	7.893 (6.3)	8.047 (5.0)
β'	1.705	1.650	1.685	1.640	α	4.619	4.574
γ	1.300	1.295	1.300	1.300	β	2.968	2.97
γ'	1.360	1.390	1.360	1.310	β'	3.094	3.11
δ	1.620	1.624	1.620	1.610	Ar	7.20	7.20 (7.0)
ε	2.960	2.943	2.950	2.950	Ar	7.22	7.28
TyrN$^\alpha$H	8.278	8.320 (6.8)	8.305 (-)	8.255 (7.5)	LysN$^\alpha$H	7.857 (5.9)	8.110 (5.0)
α	4.572	4.525 (5.6, 5.4)	4.520	4.510	α	4.180	4.200
β	2.908	2.880 (14.2)	2.880	3.060	β	1.804	1.74
β'	2.908	3.070 (14.2)	3.070	2.890	β'	1.804	1.65
γ	7.112 (8.1)	7.120 (8.2)	7.075 (7.9)	7.120 (8.0)	γ	1.40	1.41
δ	6.810	6.810	6.778	6.810	γ'	1.35	1.32
AcmCH$_3$		1.995	1.995	1.995	δ	1.676	1.63
NH		8.500	8.500	8.500	ε	3.000	2.94
CH$_2$		4.225	4.225	4.225	HisN$^\alpha$H	8.204 (6.3)	8.355 (6.5)
CH$_2$		4.275	4.275	4.275	α	4.669	4.575
					β	3.146	3.062
					β'	3.254	3.130
					δ	8.613	8.530
					ε	7.230	7.130

AlaN$^\alpha$H	8.191 (5.9)	8.30 (5.0)
α	4.370	4.293
β	1.335	1.290 (7.0)
TyrN$^\alpha$H	8.143 (5.6)	7.888 (7.0)
α	4.499	4.449
β	3.109	3.090
β'	3.109	2.920
δ	7.095 (8.0)	7.120 (8.0)
ε	6.785	6.780

building block **13** (*11*) for the synthesis of the N-glycosylated heptadecapeptide T-cell determinant from hen egg lysozyme (81-96)-Tyr, **14**. The ^{1}H NMR-spectra of the purified glycopeptide were fully assigned as presented in Tables 1 and 2.

Preliminary immunization studies in which Balb/k mice were immunized with the glycosylated MP7 (**10**) showed high antibody titers against the carbohydrate moiety of **10** Figure 1. Monoclonal antibodies against the carbohydrate are currently being generated for further characterization. All the glycopeptides **7, 8, 10** and **14** are currently under investigation to establish whether the oligosaccharide can form a part of the T-cells epitope recognized by the T-cell receptor, thus conferring T-cell specificity towards the carbohydrate.

EXPERIMENTAL

General procedures. HPLC grade solvents were purchased from Labscan Ltd. Pyridine was dried over KOH and distilled. Dichloromethane was dried over phosphorous pentaoxide and distilled. DMF was distilled by fractional vacuum distillation at 45° on a column of Rashig rings. Other chemicals were purchased as follows and used without further purification. DMAP, Dhbt-OH, Pfp-OH, DCCI and Fmoc-Su from Fluka, acetic anhydride and TFA from Aldrich, Termamyl from Novo Industries and

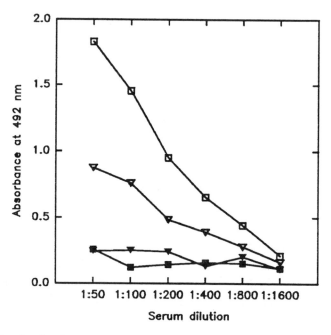

Figure 1. Antibody response towards MP7, **9** (filled symbols) and glycosylated MP7, **10** (open symbols) in two different Balb/k mice injected 4 times with glycosylated MP7. No antibodies towards the two peptides were found in non-immunized mice. A significant antibody response against MP7 itself could be detected during further immunization of the mice (data not shown).

Macrosorb SPR 100 peptide synthesis resin from Sterling Organics. The resin was derivatized with ethylene diamine, Fmoc-Nle-O-Pfp and after removal of the Fmoc group with 4-hydroxymethylphenoxyacetic acid Dhbt ester. Dhbt esters were prepared as previously described (*20*). Methoxide solution (1N) was prepared from sodium and dry methanol and diluted prior to use. HPLC was carried out on a Waters Delta Prep 3000 HPLC system with a E600 pump, a 991 photodiode array detector and preparative and analytical radial pack RP-C_{18} columns. The eluents were A: 0.1% aqueous TFA and B: 0.1% TFA in 90 % aqueous acetonitrile. Medium pressure chromatography was carried out on dry packed silica gel 60 from Merck with appropriate mixtures of ethyl acetate and petroleum ether. Silica coated aluminum TLC plates were eluted with mixtures of ethyl acetate and petroleum ether and developed by charring with sulfuric acid spray. NMR-spectra were recorded at 300°K on Bruker instruments at 500 MHz or 600 MHz. Peptide samples for NMR were prepared by dissolving the peptide in 10% D_2O in H_2O at pH 3.5 and bubbling with argon for 5 min. δ-Values are given in ppm and J-values in Hz. Microelemental analysis was carried out on a Carlo-Erba model EA1108 CHNS-O elemental analyzer. Mass spectra of compounds in a matrix of nitrobenzyl alcohol were recorded by LSIMS on a VEGA 70SE instrument with a Cs ion source calibrated with cesium iodide. Amino acid analyses of samples hydrolyzed at 110° for 24 h were carried out on an LKB alpha plus amino acid analyzer. Optical rotations were recorded on a Perkin Elmer 140 instrument.

Enzymatic degradation of amylopectin, preparation of the pentasaccharide 2. Amylopectin (40 g) was dispensed in 0.0043 M $CaCl_2$ (200 mL) at pH 5.7, and the enzyme, Termamyl 60L (0.3 g, 18 KNU) was added and the mixture incubated for 18 h at 60 °.

The product was dialyzed and the dialysate concentrated to 100 mL. Biogel P2 chromatography was performed at 65 ° on a preparative column (8×100 cm) with a flow rate of 200 mL/h and detection of fractions with a refractive index detector. Samples of 25 mL (about 5 g of oligosaccharide mixture) were applied to the column per run and four runs gave about 1 g of the branched pentasaccharide, **2** as previously characterized (*24*).

Peracetylation of compound 2. The isolated pentasaccharide **2** (840 mg, 1.00 mmol) was dispensed in pyridine (5 mL) and acetic anhydride(2.6 mL, 25 mmol) was added. The mixture was stirred and after 15 min DMAP (30 mg, 0.25 mmol) was added. The reaction was stirred overnight and the mixture was repeatedly concentrated at 0.1 torr with addition of toluene. The residue was dissolved in dichloromethane (15 mL) and washed with water (10 mL), HCl (0.1 M, 10 mL), NaHCO$_3$ solution (10 mL) and water (10 mL). The organic phase was dried with MgSO$_4$ and filtered. Concentration in vacuo yielded the anomeric mixture (1540 mg, 100%, α/β = 1/4,) of peracetylated 2, (3) which did not separate in different selected solvent mixtures in TLC.

Reaction of 3 with titanium tetrabromide. The anomeric mixture of acetates 3 (1460 mg, 0.95 mmol) was dissolved in dry chloroform (50 mL) which had been passed through basic alumina. Molecular sieves (0.5 g, 3 Å) were added and the mixture was stirred for 1 h under argon. Titanium tetrabromide (1104 mg, 3.0 mmol) was added and the mixture was refluxed under argon for 18 h until the ratio between α-acetate and glycosyl bromide remained constant according to TLC. The mixture was filtered through celite and ice water (10 mL) was added. The organic phase was separated and washed with NaHCO$_3$ solution and water. Concentration in vacuo after drying (MgSO$_4$) yielded the glycosyl bromide, 4 (1377 mg, 93%) containing 15% of unreacted α-acetate. Compound 4 was characterized by its ^{1}H-NMR data given in Table 1.

Synthesis of building block, 5, by glycosylation of Fmoc-Ser-O-Pfp with 4.
Fmoc-Ser-O-Pfp (1) (211 mg, 0.427 mmol), prepared by DCCI promoted condensation of Fmoc-Ser with Pfp-OH, purification by MPLC under dry conditions and lyophilization, was dissolved in dichloromethane (10 mL). Lyophilized bromide 4 (668 mg, 0.363 mmol) was added and the mixture was stirred over molecular sieves (.5 g, 3 Å) for 2 h at -30 °. Lyophilized silver trifluoromethanesulfonate (143 mg, 0.56 mmol) was added and the reaction was continued for 48 h at -20°. The violet product mixture was neutralized by addition of tetramethyl urea (116 μl, 1 mmol) and filtered through celite. The filtrate was washed with water and concentrated. MPLC chromatography yielded 301 mg (42 %) of the building block 5 and a small amount of the glycosylated free acid (86 mg, 13 %) as amorphous solids. Compound 5 had $[\alpha]_D^{25}$ = 72.9° (c=1 in CDCl$_3$). Anal. calc. for C$_{86}$H$_{98}$F$_5$NO$_{46}$ C, 52.26; H, 4.99; N, 0.71. Found C, 52.26; H, 4.34; N, 1.11. FAB-MS: Calc. M= 1975.52, found M^{+1}= 1976.7, M^{+23}= 1998.9. ^{1}H-NMR data for the oligosaccharide are presented in Table 2. δ/ppm (J/Hz) in CDCl$_3$: Fmoc-: 3.93, 4.03, 4.06 (CH$_2$-CH); 7.36(2H), 7.45, 7.46, 7.64, 7.66, 7.82, 7.83 (Aromatic protons); Ser: 5.785 (8.6, N$^\alpha$H), 4.840 (H$^\alpha$), 4.523 (9.3,

13.8, H$^\beta$), 4.563 (-, H$^{\beta'}$). When a ^{13}C-NMR spectrum was recorded with a delay of 3.5 s the resonances of the pentafluorophenyl ring (J$_1$ ~240 Hz) were observed at 124.81, 137.90, 140.05, 141.05, 141.05 and 142.90 ppm.

Solid phase synthesis of T-cell determinants 6 and 9 and glycosylated analogues 7, 8, 10 and 14. Macrosorb SPR 100 HMPA (3 g) was derivatized with Fmoc-Tyr(tBu)-OH (2 eqv.) according to a procedure described previously (*28*). The incorporation of tyrosine was estimated by quantitative amino acid analysis to be 62%. The dry resin (450 mg, 31 mmol) was packed in a column in our custom made fully automated solid phase synthesizer (*29*). The appropriate Fmoc-amino acid Dhbt esters (0.95 mmol each) corresponding to the peptide sequence were packed in the solid sampler vials. The glycosylated amino acids building blocks, 5 or 13 were used with addition of solid Dhbt-OH (14 mg) to the vial. The acylation reaction times in recirculating DMF were recorded by a solid phase monitor and deprotections were carried out by a 40- or a 70 min treatment with 50% morpholine in DMF. In the case of **8** these conditions were insufficient for complete deprotection of Ala-2 and Leu-3 and 4. The peptide resins were treated twice with 95% aqueous TFA (10 mL) for 1 h. After filtration and washing with 95% aqueous acetic acid the acid was removed by concentration in vacuo and the crude material was precipitated with diethyl ether. The peptides (except **14** which was too insoluble) were purified by preparative HPLC with a 1 h gradient of 20-60% buffer B in A. The yields of the acetylated glycopeptides were for **7**, 28 mg(36%); **8**, 12 mg(15%); **10**, 17 mg from 140 mg resin(65%); **14**, 26 mg(45%). The carbohydrate moiety was deprotected by suspending the insoluble peptides in methanol (10 mL) and adjusting the pH to 11 on moist pH paper. The mixture was stirred at 20° until the reaction was complete according to HPLC and then neutralized with solid CO_2. The deprotected peptides were purified by preparative HPLC with a 1 h gradient 0 - 60% buffer B in A. In the case of peptides **8** and **14** the insoluble character of these compounds caused great difficulties and some loss of material. The partially deprotected peptides still carrying Acm protection of the cysteine were isolated in 10 - 12 mg yield for all the glycopeptides which were fully characterized by 1- and 2D-^{1}H-NMR spectroscopy (*30*)as shown in Tables 1 and 2. Prior to biological studies the Acm-group was cleaved in 95% TFA (2 mL) by addition of mercury(II) acetate (10 eqv.). The volume was reduced in vacuo and the peptide was precipitated with diethyl ether which was decanted several times. The residue was dissolved in 30 % aqueous DMF treated for a period of 2 h with DTT (4 mg) at pH 4.0 and purified by HPLC as above to afford a 92 - 96% yield of peptide eluting as a single peak.

Balb/k mice were immunized weekly by subcutaneous injections of a water in oil emulsion (0.1 mL) containing equal amounts of peptide dissolved in phosphate buffered saline (PBS) and Freunds complete adjuvant. In the first injection complete adjuvant was used and incomplete adjuvant was used in the subsequent 3 injections. MP7, **9** (100 μg) or the glycosylated analog, **10** (100 μg) was injected in each animal. Sera from the mice taken before and after the immunization were analyzed with ELISA technique as previously described (*31*).

Abbreviations used. BSA, bovine serum albumine; T-cell, thymus derived lymphocyte; Pfp-, pentafluorophenyl; Dhbt-, 4-oxo-1,2,3-benzotriazin-3-yl; DMF, N,N-dimethylformamide; Fmoc-, 9-fluorenyl methoxycarbonyl; DCU, dicyclohexyl urea; HPLC, high pressure liquid chromatography; MPLC, medium pressure liquid chromatography; VLC, vacuum liquid chromatography; MHC, major histocompatibility complex; Su, succinimide; THF, tetrahydrofurane; DMAP, N,N-dimethyl aminopyridine; DCCI, dicyclohexyl carbodiimide; TFA, trifluoroacetic acid; TLC, thin layer chromatography; HMPA, hydroxymethyl phenoxyacetic acid; Acm-, acetamidomethyl; DTT, 1,4-dithio-threitol; ELISA, enzyme linked immuno sorbent assay.

Acknowledgement. This work was supported by the Technical Research Council of Denmark. We thank Dr. Carl Erik Olsen, Dept. of Chemistry, The Royal Veterinary and Agricultural University of Denmark for the recording of mass spectra.

REFERENCES

1. Harding C. V.; Roof, R. W.; Allen, P. W.; Unanue, E.R. *Proc. Natl. Acad. Sci. USA* **1991**, 88, 2740.
2. Lavielle, S.; Ling, N. C.; Guillemin, R. C. *Carbohydr. Res.* **1981**, 89, 221.
3. Kunz, H.; B. Dombo, B. *Angew. Chem., Int. Ed. Engl.* **1988**, 27, 711.
4. Paulsen, H.; Merz, G.; Weichert, U. *Angew. Chem., Int. Ed. Engl.* **1988**, 27, 1365.
5. Lüning, B.; Norberg, T.; Tejbrant, J. *J. Chem. Soc., Chem. Commun.* **1989**, 1267.

6. Filira, F.; Biondi, L.; Scolaro, B.; Foffani, M. T.; Mammi, S.; Peggion, E.; Rocchi, R. *Int. J. Biol. Macromol.* **1990**, 12, 41.

7. Bardaji, E.; Torres, J. L.; Clapés, P.; Albericio, F.; Barany, G.; Valencia G. *Angew. Chem., Int. Ed. Engl.* **1990**, 29, 291.

8. Filira, F.; Biondi, L.; Cavaggion, F.; Scolaro, B.; Rocchi, R. *Int. J. Pept. Protein Res.* **1990**, 36, 86.

9. Meldal, M.; Jensen, K. J. *J. Chem. Soc., Chem. Commun.* **1990**, 483.

10. Paulsen, H.; Merz, G.; Peters, S.; Weichert, U. *Liebigs Ann. Chem.* **1990**, 1165.

11. Meldal, M. and Bock, K. *Tetrahedron Lett.* **1990**, 48, 6987.

12. Otvos, L., Jr.; Urge, L.; Hollosi, U.; Wroblewski, K.; Graczyk, G.; Fasman, G. D.; Thurin, J. *Tetrahedron Lett.* **1990**, 31, 5889.

13. Jansson, A. M.; Meldal, M.; Bock, K. *Tetrahedron Lett.* **1990**, 48, 6991.

14. Urge, L.; Kollat, E.; Hollosi, M.; Laczko, I.; Wroblewski, K.; Thurin, J.; Otvos L., Jr. *Tetrahedron Lett.* **1991**, 32, 3445.

15. Peters, S.; Bielfeldt, T.; Meldal, M.; Bock, K.; Paulsen, H. *Tetrahedron Lett.* **1991**, 32, 5067.

16. Schultheiss-Reimann, P.; Kunz, H. *Angew. Chem., Int. Ed. Engl.* **1983**, 22, 62.

17. Waldmann, H.; März, J.; Kunz, H. *Carbohydr. Res.* **1989**, 196, 75.

18. Shaban M. A. E.; Jeanloz, R. W. *Bull. Chem. Soc. Jpn.* **1981**, 54, 3570.

19. Lavielle, S.; Ling, N. C.; Guillemin, R. C. *Carbohydr.Res.* **1981**, 89, 221.

20. Atherton, E.; Holder, J. L.; Meldal, M.; Sheppard, R. C.; Valerio, R. M. *J. Chem. Soc., Perkin Trans. 1* **1988**, 2887.

21. Kisfaludy, L.; Schön, I. *Synthesis* **1983**, 325.

22. Schön, I.; Kisfaludy, L. *Synthesis* **1986**, 303.

23. Jensen, K. J.; Meldal, M.; Bock. K. *Peptides, Proc. 12'th Am. Pep. Symp.* **1991**, In Press.

24. Bock, K. *Pure & Appl. Chem.* **1987**, 59, 1447.

25. Mouritsen, S.; Meldal, M.; Rubin, B.; Holm, A.; Werdelin, O. *Scand. J. Immunol.* **1989**, 30, 723.

26. Pardi, A.; Billeter, M.; Wüthrich, K. *J. Mol. Biol.* **1984**, 180, 741.

27. Shenderovich, M. D.; Nikiforovich, G. V.; Chipens, G. I.. *Magn. Reson.* **1984**, 59, 1.

28. Blankemeyer-Menge, B.; Frank, R.; *Tetrahedron Lett.* **1988**, 29, 5871.

29. Cameron, L. R.; Holder, J. L.; Meldal, M.; Sheppard, R. C. *J. Chem. Soc. Perkin Trans. 1.* **1988**, 2895.

30. Bock, K.; Duus, J. Ø.; Norman, B.; Pedersen, S. *Carbohydr. Res.*,**1991**, 211, 219.

31. Mouritsen, S.; Werdelin, O.; Meldal, M.; Holm, A. *Scand. J. Immunol.* **1988**, 28, 261.

RECEIVED April 15, 1992

Chapter 4

Human Fucosyltransferases

Involved in the Biosynthesis of X (Gal-β-1—4[Fuc-α-1—3]GlcNAc) and Sialyl—X (NeuAC-α-2—3Gal-β-1—4[Fuc-α-1—3]GlcNAc) Antigenic Determinants

Winifred M. Watkins[1], Patricia O. Skacel[2], and Philip H. Johnson[3]

[1]Department of Haematology, Royal Postgraduate Medical School, Hammersmith Hospital, London W12 ONN, England
[2]Department of Haematology, Northwick Park Hospital, Watford Road, Harrow, Middlesex HA1 3UJ, England
[3]Medical Research Council Human Biochemical Genetics Unit, University College, London NW1 2HE, England

The X structure (Gal-β-1-4[Fuc-α-1-3]GlcNAc) occurs in glycoproteins, glycolipids, and free oligosaccharides and may be exposed as a terminal non-reducing end-group or masked by substitution with other sugars. The presence of sialic acid in α-2,3-linkage to the terminal β-galactosyl residue gives the sialyl-X determinant (NeuAc-α-2-3Gal-β-1-4-[Fuc-α-1-3]-GlcNAc). Both structures have been identified as human tumour markers and as ligands in cellular adhesion reactions mediated by endogenous lectins (LEC-CAMS). X structures are biosynthesised by a family of GDP-fucose:-N-acetyl-D-glucosaminide α-3-L-fucosyltransferases. The obligatory pathway for biosynthesis of sialyl-X is first the formation of NeuAc α-2-3-Gal-β-1-4GlcNAc, followed by the addition of the fucosyl residue to the N-acetylglucosamine unit. Not all α-3-fucosyltransferase species that synthesise X determinants can add fucose to the GlcNAc residue in NeuAc α-2-3Gal-β-1-4-GlcNAc to form sialyl-X determinants. In myeloid tissues species differing in their activities with sialylated acceptors occur at different stages of maturation; these variants appear to be encoded by different genes from the α-3-fucosyltransferases expressed in other tissues.

0097–6156/93/0519–0034$08.50/0

The carbohydrate moieties of glycoproteins, glycosphingolipids and glycosaminoglycans are involved in many cell
surface phenomena, including antibody and lectin binding,
cellular adhesion and recognition. Changes that take
place in the surface carbohydrate structures are thought
to play an essential part in normal cellular differentiation and to influence the metastatic potential of cancer
cells. The possibility of studying these changes has been
greatly facilitated by the advent of the hybridoma
technique (1) which has allowed the production of a range
of highly specific monoclonal antibodies for the detection of closely related carbohydrate structures (2,3). Of
equal importance for an understanding of the biochemical
and genetic mechanisms underlying the changing pattern of
cell surface structures has been the characterisation of
the glycosyltransferase enzymes (reviewed in refs 4 & 5)
responsible for the biosynthesis of the determinants.
Assay of these glycosyltransferases enables attempts to
be made to correlate surface antigenic changes with loss
or overproduction of the enzymes, and the recent success
in cloning glycosyltransferase genes (6-8) is already
providing valuable probes for further study of the genetic regulation and tissue specific expression of the enzymes and antigens.
      L-Fucose and sialic acid (N-acetylneuraminic acid)
are sugars involved in the terminal glycosylation of
oligosaccharide chains of glycoconjugates and as such
play a major role in the antigenic properties of cell
surfaces and in the changing carbohydrate profiles that
occur on cell maturation and in malignancy. This communication will discuss 1) some of the antigenic structures
involving fucose and sialic acid, with an emphasis on X
and sialyl-X (Figure 1), 2) the roles of these structures
as antigenic determinants, tumour markers and ligands for
adhesion molecules and 3) the properties and interrelationships of the fucosyltransferases responsible for the
biosynthesis of X and sialyl-X.

Fucosyl-linkages in Human Glycoconjugates

Four different types of fucosyl-linkages are commonly
found in human glycoconjugates (Figure 2). The first two
to be clearly associated with antigenic determinants were
fucose linked α-1,4 to a subterminal N-acetylglucosaminyl residue in a Type 1 (Gal-β-1-3GlcNAc) chain to give
the Le^a antigenic determinant (9) and fucose linked α-
1,2 to a β-galactosyl residue in either a Type 1 or a
Type 2 (Gal-β-1-4GlcNAc) oligosaccharide chain ending, to
give blood group H antigenic structures (10).
      Subsequently the presence of both α-1,2-linked and
1,4-linked fucose in a Type 1 chain structure (Fuc-α-1-
2Gal-β-1-3[Fuc-α-1-4]GlcNAc) was found to be responsible
for Le^b antigenic specificity (11). Characterisation of
this determinant showed that the addition of a monosaccharide substituent to a sugar residue adjacent to a determinant structure in an oligosaccharide chain, whether
distal or proximal to the first substituent, masks the

STRUCTURE	REFERENCE

X-determinant

Gal β 1-4
$\qquad\qquad$ GlcNAc- β -R $\qquad\qquad\qquad\qquad$ 14.16
Fuc α 1-3

Sialyl-X determinant

NeuAc α 2-3Gal β 1-4
$\qquad\qquad\qquad$ GlcNAc- β -R $\qquad\qquad\qquad$ 23-27
$\qquad$ Fuc α 1-3

VIM 2 determinant

NeuAc α 2-3Gal β 1-4GlcNAc β 1-3Gal β 1-4
$\qquad\qquad\qquad\qquad\qquad\qquad\qquad\qquad$ GlcNAc- β -R $\qquad$ 28
$\qquad\qquad\qquad\qquad$ Fuc α 1-3

Dimeric-X determinant

Gal β 1-4
$\qquad\qquad$ GlcNAc β 1-3Gal β 1-4
Fuc α 1-3 $\qquad\qquad\qquad\qquad$ GlcNAc- β -R $\qquad$ 30
$\qquad\qquad\qquad\qquad$ Fuc α 1-3

Figure 1. Structures of X, Sialyl-X, VIM 2 and Dimeric-X determinants. All sugars are in the pyranose form. Abbreviations:- Gal, D-galactose; Fuc, L-fucose; GlcNAc, N-acetyl-D-glucosamine; NeuAc, N-acetylneuraminic acid R, remainder of molecule.

Linkage formed by Fucosyltransferase	Acceptor Substrate	Structure Formed
Fuc– α -2–	Gal– β –R	Fuc-α-1–2Gal– β –R
Fuc– α -3–	Gal–β-1–4GlcNAc–R (Type2)	Gal-β-1–4 $\diagdown$ GlcNAc–R / Fuc-α-1–3
Fuc– α -3–	Gal–β-1–4Glc (Lactose)	Gal-β-1–4 $\diagdown$ Glc / Fuc α 1-3
Fuc– α -4–	Gal–β-1–3GlcNAc–R (Type 1)	Gal-β-1–3 $\diagdown$ GlcNAc–R / Fuc-α-1–4
Fuc– α -6–	R...GlcNAc-β-1–4GlcNAc–ASN	R...GlcNAc-β-1–4GlcNAc–ASN 6 | 1 Fuc

Figure 2. Fucosyl linkages commonly found in human glycoconjugates. Abbreviations:- Glc, D-glucose; ASN, asparagine; others as in Figure 1.

original specificity and gives rise to a structure that
has its own distinctive antigenic properties.

In general, Type 2 H determinants occur as terminal
structures in glycosphingolipids and in glycoproteins
with both O-linked and N-linked oligosaccharide chains
and Type I determinants (H, Le^a and Le^b) in glycosphingolipids and in glycoproteins with O-linked chains
(4).

No antigenic activity has been associated with fucose
linked   α-1,6 to N-acetylglucosamine (Figure 2). This
linkage appears to occur exclusively in the core region
on the N-acetylglucosamine residue attached to asparagine
in the peptide backbone of glycoproteins with N-linked
chains (12).

Fucose linked α-1,3 to N-acetylglucosamine was first
reported in oligosaccharides obtained by degradation of
O-linked chains in blood-group-active ovarian cyst glycoproteins. Lloyd et al (13) obtained fragments from A- and
H-active glycoproteins that contained internal linkages
of fucose joined α-1,3 to N-acetylglucosamine. The trisaccharide sequence, Gal-β-1-4[Fucα-1-3]GlcNAc, was first
isolated in 1968 by Marr et al.(14) as a terminal non-
reducing structure from a glycoprotein preparation obtained from the cyst fluid of an Le^a-positive donor.
Since ovarian cyst glycoproteins have both Type 1 and
Type 2 oligosaccharide chains (15), preparations obtained
from Le^a-positive donors, who are ABH-non-secretors and
hence lack terminal A-, B-, or H-active structures, can
yield oligosaccharide fragments in which the terminal
non-reducing end groups of Type 1 chains have, as the
only substituent, fucose in α-1,4 linkage to the subterminal N-acetylglucosamine (Le^a determinant) and Type 2
chains have, also as the only substituent, fucose linked
α-1,3 to N-acetylglucosamine (X determinant).The
isolated trisaccharide was recognised as an isomer of Le^a
that was without Le^a immunological activity (14).

The name X-hapten was subsequently applied to the
Type 2 trisaccharide with α -3-linked fucose when a
glycosphingolipid with this terminal non-reducing sequence was isolated from human adenocarcinoma tissue (16).
The linkage is now known to occur in a variety of glycoproteins with both O- and N-linked oligosaccharide chains
and in free lactose-based oligosaccharides in human milk
(17) and urine (18). In poly-N-acetyllactosamine chains
in glycoproteins and glycolipids fucose residues may also
be attached to the O-3-position of internal N-acetylglucosamine residues (Dimeric-X, Figure 1) (19). The
lactose-based oligosaccharides can also have fucose linked α-1-3 to the glucose residue in the lactose (Gal β1-
4Glc) structure (17-18) (Figure 2).

Partial characterisation of a human antibody that
agglutinated erythrocytes of individuals grouped as
"Le(a-b-), non-secretors of ABH" suggested that this
antibody, designated as anti-Le^c, might be detecting the
X-haptenic sequence (20). No other naturally occurring

human antibodies reacting with this structure appear to
have been reported. With the introduction of the hybridoma technique for the production of monoclonal antibodies, however, the Gal-β-1-4[Fuc-α-1-3]GlcNAc structure
was found to be strongly immunogenic in mice and antibodies raised against a variety of cell lines, as well as
those raised against both normal and tumour tissues, were
subsequently found to recognise this trisaccharide sequence (2-3,21). Among those frequently referred to in the
literature are an antibody directed to a stage specific
embryonic antigen in the mouse (SSEA 1, ref.22) and a
group of human leukocyte antibodies called CD 15 (Cluster
of Differentiation 15; International Workshop on
Differentiation Antigens, 1987). The symbols Le^x or
Lewis-x are also used for the trisaccharide sequence and
the difucosyl structure (Fuc-α-1-2Gal-β 1-4[Fuc-α-1-3]-
GlcNAc), the Type 2 chain isomer of Le^b, is known as Y,
Le^y, or Lewis-y.

Hybrid Structures containing Fucose and Sialic acid

The Type 1 and Type 2 structures in glycoproteins and
glycolipids are substrates not only for the fucosyltransferases but also for the enzymes that transfer N-
acetylneuraminic acid (sialic acid) to the terminal β-
galactosyl residues to form NeuAc 2-3Gal and NeuAc 2-6-
Gal linkages (5). Glycoconjugate molecules bearing sialic
acid and fucose joined to adjacent sugar residues (Figure
1) have been isolated from both normal human tissues and
from tumours (23-27). When the sialic acid is linked α-
2,3 on the terminal β-galactosyl residue the structure
formed on Type 1 chains is sialyl-Le^a and on Type 2
chains is sialyl-X (or sialyl-Le^x). Monoclonal antibodies
specific for these hybrid structures have been raised by
immunisation of mice with cancer cell membranes or with
purified glycolipids antigens coated onto bacteria
(25,27). The presence of the sialic acid residue masks
the specificity of the Le^a or X structures in the same
way as the presence of a fucosyl residue on the terminal
galactose masks the specificity of Le^a in Le^b structures
and X in Y structures.

    An additional structure carrying a terminal α-2,3-
linked sialic acid and a fucose on an internal N-
acetylglucosamine residue of a poly-N-acetyllactosamine
chain (Figure 1) is recognised as an independent antigen
by the monoclonal antibody VIM 2 (28).

X and Sialyl-X as Tumour-Associated and Differentiation Antigens

Although X and sialyl-X structures have been identified
in glycoconjugates from normal human tissues both occur
in higher density as surface components in certain tumour
tissues. Glycosphingolipids carrying the X determinant

were found to accumulate in human lung, gastric and colonic cancers (16,29) and compounds with poly-N-acetyllactosamine chains have been isolated in which two
(Figure 1) and three fucose residues are joined in   α-
3-linkage to successive internal N-acetylglucosamine residues (30). In addition many human cancers, particularly
adenocarcinomas, have been characterised by the presence
of the sialyl-X antigen in organs in which it is either
not normally expressed, or only to a much lesser extent
(25,30-31). The role of X and sialyl-X as tumour markers
has suggested the use of monoclonal antibodies directed
towards these determinants as therapeutic reagents for
drug targeting and tumour suppression (29).
     In human haemopoietic tissue X antigenic expression
appears to be developmentally regulated since it is found
on mature neutrophils and on all myeloid cells from the
promyelocytic stage onwards but is absent from earlier
precursors (32-33). The X structure was first recognised
as a developmentally regulated antigen in the mouse inas-
much as a monoclonal antibody directed to the undiffer-
entiated teratocarcinoma cell line F9 (anti-SSEA 1, ref.
22), which was shown to be reacting with the X sequence
(34), was found to be maximally expressed at the morula
to early blastocyst stages and to decline at the later
stages of development. In human tissues X and Dimeric-X
are expressed in gastrointestinal tissue early in embryogenesis but virtually disappear from the mucosa of new-
born infants and adults, only to reappear in gastrointestinal tumours (30).

**The Role of X and Sialyl-X as Ligands for Cell Adhesion
Molecules**

The interest generated by the finding that X and sialyl-X
are tumour and differentiation markers has recently been
far exceeded by interest in the finding that these
carbohydrate structures possibly function as ligands for
the family of adhesion molecules known as the LEC-CAMS
(or Selectins). Adhesion of circulating leukocytes to
vascular endothelial cells is a key step in inflammatory
responses and the LEC-CAMs have been implicated in the
interaction of leukocytes with platelets or vascular
endothelium (35-36). These proteins are a unique family
of cell adhesion molecules characterised by the juxta-
position of an N-terminal lectin domain, an epidermal
growth factor domain, and variable numbers of complement
regulatory protein-like repeating units (reviewed in refs
37 & 38). The minimal oligosaccharide bound by LEC-CAM 1,
which is constitutively expressed on the majority of
leukocytes and facilitates binding of these cells to
endothelium during lymphocyte recirculation, has not yet
been clearly defined but sialic acid or other charged
groups have been implicated (39).
     The ligand most firmly established appears to be

that recognised by LEC-CAM 2 (also known as ELAM 1) which
is transiently expressed by endothelial cells in response
to inflammatory agents. Several groups have concluded
that this adhesion molecule recognises the sialyl-X
structure (40-42). An alternative suggestion (43) that
this LEC-CAM recognises the VIM 2 determinant (Figure 1),
has not been confirmed by recent work (44).
     LEC-CAM 3 the third member of this family (also known
as PADGEM, GMP 140 or CD62) is found in platelets and
endothelial cells and binds to neutrophils and monocytes
at the site of tissue injury. Antibody blocking experiments and the use of soluble inhibitors strongly indicated that the X-haptenic structure, or a closely related
oligosaccharide sequence, is involved in the binding of
this adhesion molecule (45).
     The demonstration of the role of sialyl-X and
related carbohydrate structures as ligands for adhesion
molecules has suggested new possibilities for the development of therapeutic agents for treatment of inflammatory conditions and appears at last to have brought
glycoconjugates into the forefront of biochemical and
biomedical thought (37-38, 44,46).

Biosynthesis of X and Sialyl-X Structures

**Is the X Structure Biosynthesised by the Lewis Gene-
Associated α-3/4-Fucosyltransferase?** GDP-L-fucose:-
galactosyl-β-1-4-N-acetyl- β-D-glucosaminide α-3-fucosyl-
transferases able to transfer fucose to the O-3-position
of the penultimate N-acetylglucosamine in Type 2 structures to form the X-determinant were early detected in
human milk, submaxillary glands, stomach mucosa and
plasma (47-49). Although all the tissues expressing the
Lewis Le gene associated GDP-L-fucose:galactosyl β-1-3-N-
acetyl $-$ β-D-glucosaminide α-4-fucosyltransferase also
have 3-fucosyltransferase activity not all the tissues
with α-3-activity have α-4-activity. This fact, together with the finding that 3-fucosyltransferase activity is expressed in tissues of individuals of the
genotype Le(a-b-), indicated that the α-3-enzyme is encoded by a gene independent of the Lewis gene (4). The
differences in tissue specific expression of the α-3- and
the α-4-enzymes, together with the known frequency of
individuals homozygous for the Le gene (50), also suggested that the 3-fucosyltransferase is not the product of
the second le allele at the Lewis locus (4).
     Subsequently examination of saliva from individuals
of different Lewis groups for fucosyltransferase activities demonstrated that GDP-L-fucose:galactosyl-β -1-4-
D-glucose α -3-fucosyltransferase activity is expressed
only when the individual has a Lewis gene (51). The co-
presence of the two enzymes meant that it was not possible from these experiments with whole saliva to deduce
whether the Lewis enzyme could transfer fucose to the O-

3-, as well as to the O-4-, position of N-acetylglucosamine. The capacity of a highly purified fucosyltransferase from human milk to transfer L-fucose to N-acetylglucosamine in both Type 1 and Type 2 chains led Prieels
et al.(52) to conclude that the Le-gene encoded enzyme is
an α-3/4-fucosyltransferase that can synthesise both X
and Le^a structures. In our laboratory, however, further
fractionation on Sephacryl S-200 of a similarly purified
preparation of α-3/4-fucosyltransferase from human milk
removed a large proportion of the enzyme activity
responsible for the transfer of L-fucose to the O-3-
position of the subterminal N-acetylglucosamine residue
in low molecular weight Type 2 oligosaccharides; the
resultant enzyme reacted mainly with N-acetylglucosamine
in Type 1 chains and D-glucose in lactose-based structures and had no detectable capacity to transfer fucose to
Type 2 chains in glycoproteins (Table I) (53-54). A
similarly purified 3/4-fucosyltransferase from A431 cells
behaved in the same way towards Type 2 substrates (Table
I) (55).
     The acceptor specificity of the further purified
Lewis 3/4-fucosyltransferase therefore suggested that the
Le-gene encoded enzyme would not have the capacity to
synthesise X-haptenic structures to any appreciable
extent in vivo. However, Kukowska-Latallo et al.(8)
recently cloned the Le gene from the A431 cell line using
a mammalian gene transfer procedure and demonstrated that
COS-1 cells transfected with this gene 1) express both
Le^a and X determinants that were not expressed by the un-
transfected cell and 2) contained a fucosyltransferase
that reacted with both Type 1 and Type 2 substrates. This
apparent discrepancy remains to be resolved. One possible
explanation is that the soluble enzyme in milk has been
digested by proteases to release it from the Golgi membranes and this process has resulted in some conformational change in the enzyme protein that influences its
fine specificity.

Pathway of Biosynthesis of Sialyl-X Structures. The α-3/4
-fucosyltransferase preparation purified from human milk
by Prieels et al.(52) was said not to transfer fucose to
sialic acid-containing substrates and from this result
the pathway of biosynthesis of sialyl-X was un-clear
since it had previously been shown that the X (Gal-β-1-4-
[Fuc-α-1-3]GlcNAc) structure is not a substrate for the
6'- or 3'-sialyltransferases (56). However, Johnson et
al.(57), using a different procedure for assay and
identification of the product, found that partially purified 3/4-fucosyltransferase from human milk could transfer fucose to NeuAc α-2-3Gal-β1-4GlcNAc to form sialyl-X,
although NeuAc-α-2-6Gal-β-1-4GlcNAc was not a substrate.
The 3-fucosyltransferase in human milk, separable from
the 3/4-fucosyltransferase (58), the corresponding enz-

ymes purified from plasma and liver (57,59) (Table II),
the enzyme in human lung carcinoma cells (27) and an 3-
fucosyltransferase in human amniotic fluid (60) were also
shown to utilise compounds with terminal NeuAc-α-2-3Gal-
β 1-4GlcNAc sequences to form sialyl-X structures (Figure
3). More recently Mollicone et al.(61) re-examined homogenates of various tissues for 3-fucosyltransferase
activity and confirmed the utilisation of sialyl acceptors by the α-3-enzymes in plasma and liver and by the
3/4-enzymes in milk, kidney, and gall bladder.
    The pattern of biosynthesis of sialyl-X thus resemb-
les the formation of the difucosyl Le^b and Y structures
insofar as in these determinants the addition of the α-
2-linked fucose to the terminal galactose residue must
precede the transfer of the second fucose unit to the
subterminal N-acetylglucosaminyl unit (4-5). Presumably
when fucose is transferred to the N-acetylglucosamine
residue in an unsubstituted Type 1 or Type 2 disaccharide
structure the galactose is either shielded from access by
the enzymes that normally glycosylate the terminal sugar
or the conformation is changed is such a way as to make
the galactose unrecognisable by the enzymes. Although the
α-3/4-, and certain α-3-fucosyltransferases, add fucose
to the first internal N-acetylglucosamine unit when the
adjacent β-galactosyl residue is subterminal to another
sugar, addition of fucose to the N-acetylglucosamine residue before the addition of the substituent on the galactose, with the formation of Le^a or X structures,
precludes further elongation of oligosaccharide chains.

**Evidence for Multiple GDP-Fucose:Galactosyl β1-4-N-
Acetylglucosaminide α-3-Fucosyltransferases**

**Polymorphism of the Gene Encoding the Serum 3-Fucosyl-
Transferase Activity.** In the Caucasian population some 6%
of individuals are grouped as Le(a-b-) on the basis of
red cell and saliva tests (50) and these persons fail to
express the 4-fucosyltransferase in tissues, such as
milk, saliva and intestinal mucosa, in which the enzyme
is found in Lewis positive individuals (62,47,48).
Examination of many serum and plasma samples for the GDP-
fucose:galactosyl-β-1-4-N-acetyl— β—D-glucosaminide α-3-
fucosyltransferase indicated that if there is genetic
polymorphism at the locus encoding this enzyme the frequency of an inactive allele must be very low. However,
in 1986 two unrelated individuals were found to virtually
lack 3-fucosyltransferase activity in their serum and
saliva (63); both were of the Lewis blood group phenotype Le(a-b-) and both were members of Black families,
one from the United States and the other from South
Africa. No others of the tested members of these families
lacked this enzyme.
    At this time a survey of more than 2000 serum samp-
les from both Black and White blood donors failed to

Table I. Substrate Specificities of α-3/4-Fucosyltransfer-
ases Purified from Human Milk and from A 431 Cells

| Substrate | Relative activities[*] of α-3/4-fucosyltransferases from: | | |
| | Milk[**] | | A 431 cells |
	Before Sephacryl	After Sephacryl	
Gal-β-1-3GlcNAc	100	100	100
Gal-β-1-4GlcNAc	105	11	11
NeuAc-α-2-3Gal-β-1-4GlcNAc	88	38	15
NeuAc-α-2-6Gal-β-1-4GlcNAc	0	0	0
Fuc-α-1-2Gal-β-1-4GlcNAc	94	nt	72
Fuc-α-1-2Gal-β-1-4Glc	69	91	115
Transferrin	3	0	0
Asialo-transferrin	16	0	0
α_1-Acid glycoprotein	29	0	0
Asialo-α_1-acid glycoprotein	30	0	0

* Relative activity with 0.5 μmol low molecular weight
  acceptors and 100 μg glycoprotein acceptors.
** Purified α-3/4-fucosyltransferase from milk of
  Le(a-b+) donor further fractionated on Sephacryl S-200.
SOURCE: Compiled from data in References 55; 57; 69

Table II. Substrate Specificities of Purified Human
α-3-L-Fucosyltransferases

| Substrate | Relative activities[*] of 3-fucosyltransferases from: | | | | |
| | Plasma | Milk[**] | Liver | Neutrophils | |
				Normal	CML
Gal-β-1-4GlcNAc	100	100	100	100	100
Gal-β-1-3GlcNAc	0	0	0	0	0
NeuAc-α-2-3Gal-β-1-4GlcNAc	147	109	133	46	2
NeuAc-α-2-6Gal-β-1-4GlcNAc	0	0	0	0	0
Fuc-α-1-2Gal-β-1-4GlcNAc	162	123	120	106	163

* Incorporation (c.p.m.) of [^{14}C]fucose into 0.5 μmol
  acceptor relative to incorporation into 0.5 μmol
  Gal-β-1-4GlcNAc.
** 3-Fucosyltranferase purified from milk and free from
  Lewis-gene associated 3/4-fucosyltransferase activity
SOURCE: Compiled from data in References 57;59;69.

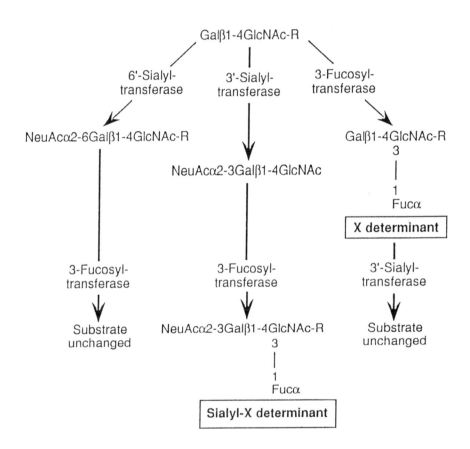

Figure 3. Biosynthetic pathways for the formation of X and sialyl-X structures. Abbreviations as in Figure 1.

disclose another example of a serum deficient in α -3-
fucosyltranferase activity (63) but further examples of
this phenotype were subsequently discovered in France in
two members of an African family (64); these two individuals also belonged to the Le(a-b-) red cell phenotype.
However, examination of their white cells revealed the
presence of X and sialyl-X antigens, indicating that,
despite the deficiency of 3-fucosyltransferase in their
sera, this enzyme activity must   still be expressed in
their haemopoietic tissues. These results therefore
pointed to the possibility that the 3-fucosyltransferase
gene responsible for the activity found in plasma might
be different from the gene encoding the enzyme in
leukocytes.
        More recently Oriol et al. (65) encountered some
eighteen sera deficient in α-3-fucosyltransferase   activity in random samples collected in Indonesia; all the
donors of these samples were also found to belong to the
Lewis negative Le(a-b-) red cell phenotype. Hence twenty-
two individuals belonging to the rare phenotype characterised by a deficiency of 3-fucosyltransferase activity in serum also failed to express 4-fucosyltrans-
ferase in epithelial tissues. Therefore, these studies
not only suggested that the 3-fucosyltransferases in
plasma might have a different genetic origin from the
enzyme in myeloid tissue but also revealed a close relationship between the plasma enzyme and the Lewis gene
associated 3/4-fucosyltransferase despite the fact that
the enzymes have different acceptor specificities and
tissue specific expression.

        **Purification of α-3-Fucosyltransferases from Normal
and Leukaemic Leukocytes.** Further evidence for the existence of more than one GDP-L-fucose:galactosylβ1-4-N-
acetyl-β-D-glucosaminide α-3-fucosyltransferase   has
come from purification of the enzymes in myeloid tissues.
In neutrophils the most abundant carbohydrate structures
on glycoproteins and glycolipids are based on Type 2
poly-N-acetyllactosamine chains composed of repeating
Gal-β-1-4-GlcNAc subunits. Both the X and the sialyl-X
structures have been reported as terminal non-reducing
groups in oligosaccharides isolated from normal human
neutrophils (66-68). The monoclonal antibody VIM-2,
characterised as recognising the sialyl-fucosyl-oligo-
saccharide structure with the fucosyl residue on an
internal N-acetylglucosamine residue (Figure 1), was
found to be specific for human blood cells of myelomono-
cytic lineage (28). The level of activity varied among
different myeloid cells and, from studies on normal and
leukaemic cells, the authors concluded that the total
amount of gangliosides bound by this antibody was related
to the level of cellular differentiation.
        Purification of the 3-fucosyltransferases from normal
neutrophils and from cells obtained from a patient with

CML yielded preparations that were similar in many of
their properties but differed strikingly in one respect;
whereas the enzyme from normal cells utilised 3'-sialyl-
N-acetyllactosamine as a substrate, with an activity
equivalent to about half that found for unsubstituted
Gal β 1-4GlcNAc, the preparation purified 200,000 fold
from the leukaemic cells had virtually no activity with
this substrate (Table II) and utilised fetuin or α_1-acid
glycoprotein to only a very limited extent unless sialic
acid had been removed from the glycoproteins (69).
      The inference from these results was that not only
does the enzyme in myeloid cells differ from that in
plasma but also there could be two species of 3-fucosyl-
transferase in myeloid cells differing in their specificity with regard to sialic acid-containing acceptors. In
addition, the 3-fucosyltransferase purified from normal
neutrophils that utilised 3'-sialyl-N-acetyllactosamine
as an acceptor differed from the enzymes purified from
plasma, milk and liver in that the activity with this
substrate, although quite definite, was less than with
the unsubstituted N-acetyllactosamine whereas with the
other enzymes the sialylated compound was the preferred
substrate (Table II).

**Antibodies Raised Against the α-3/4- and α-3-Fucosyl-
transferases.** The 3/4-fucosyltransferase purified
300,000-fold from human milk (58) and the 3-fucosyltrans-
ferase purified 100,000-fold from human liver (59) were
used to raise polyclonal rabbit antibodies. These sera
provided further evidence for a close relationship
between the Lewis gene-associated 3/4-fucosyl-transferase
and the plasma, liver and milk GDP-fucose:galactosyl β 1-
4-N-acetylglucosaminide α-3-fucosyltransferases (Figure
4). Both antibodies cross reacted with each of these
enzymes (55). In contrast, the antibody to the 3/4-
fucosyltransferase failed to react with either the 3-
fucosyltransferase isolated from normal neutrophils or
the enzyme purified from leukaemic cells (Johnson,P.H.;
Watkins W.M., unpublished data). Further characterisation
of these antibodies is required to ensure that they are
detecting protein and not carbohydrate epitopes, which
could differ according to the cellular origin of the
enzyme protein. However, at their face value, the results
indicate 1) structural homology between the Lewis-gene
associated α-3/4-fucosyltransferase and the non-myeloid
group of α-3-fucosyltranferases and 2) lack of structural homology between this group and the two different
species found in myeloid tissue.

Chromosomal Localisation of Fucosyltransferases.
Evidence that the Lewis-gene-encoded enzyme has a
different chromosomal location from at least one of the
myeloid 3-fucosyltransferases has come through linkage
studies and somatic cell hybrid experiments. The Lewis

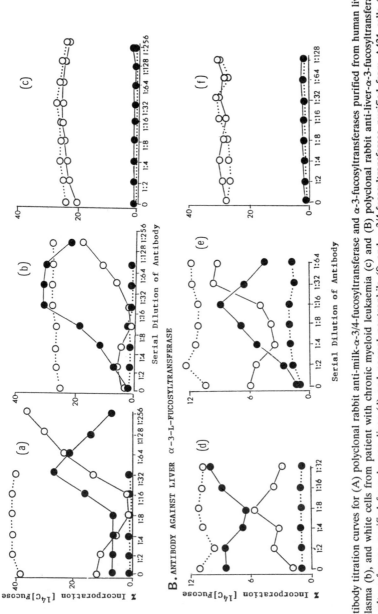

Figure 4. Antibody titration curves for (A) polyclonal rabbit anti-milk-α-3/4-fucosyltransferase and α-3-fucosyltransferase purified from human liver (a), human plasma (b), and white cells from patient with chronic myeloid leukaemia (c) and (B) polyclonal rabbit anti-liver-α-3-fucosyltransferase with α-3-fucosyltransferases purified from human liver (d) and normal neutrophils (f) and α-3/4-fucosyltransferase purified from A431 cells (e). Serial dilutions of antibody were mixed with a constant amount of enzyme preparation and the immune complexes were precipitated with a donkey anti-rabbit-serum. After centrifugation, enzyme activity with Gal-β-1-4GlcNAc (for α-3-activity) or Gal-β-1-3GlcNAc (for α-4-activity) was measured in the precipitate (●——●) and the supernatant (○——○). Preimmunization rabbit serum samples (●·····●, precipitate; ○····○, supernatant) were included as controls. (Adapted from reference 55 and unpublished data.)

blood-group locus was assigned to chromosome 19 by virtue
of its linkage to the locus for complement C3 (70) and
the cloned human cDNA encoding the α-3/4-fucosyltrans-
ferase was shown by Southern blot analysis to cross-
hybridise with sequences on chromosome 19 (8). In contrast, somatic cell hybrid experiments have led to the
assignment of at least one of the myeloid α-3-fucosyl-
transferase species to human chromosome 11 (71-73). The
chromosomal locations of the genes encoding the liver or
kidney 3-fucosyltransferases, or the species that appear
in soluble form in plasma or milk, have not yet been
established.

Susceptibility to Sulphydryl Reagents.   Chou et al.(74)
observed differences in the susceptibilty of fucosyltransferases in plasma to inhibition by N-ethylmaleimide,
although the precise specificities of the enzymes were
not clearly defined at that time. Subsequent studies
(55,61) have shown that the plasma 3-fucosyltransferase
is much more sensitive to inhibition by this sulphydryl
reagent than are the enzymes in myeloid cells. Interest-
ingly, Mollicone et al.(61) found that the enzyme in
brain resembled the leukocyte species in its resistance
to N-ethylmaleimide whereas a survey of kidney, liver,
and gall-bladder showed that the 3-fucosyltransferases in
these tissues behaved similarly to the plasma enzyme.

**Enzymic Control of the Expression of X and Sialyl-X in
Human Myeloid Cells.**

**Transferase and antigen expression in platelets and
leukocytes from healthy individuals and patients with
leukaemia.** Studies initially designed to correlate glycosyltransferase and antigenic expression in leukaemic
cells have yielded information relevant to an understanding of the changes that take place in the process of
normal myeloid cell maturation (75). Preliminary examination of fucosyltransferase activities in normal human
platelets and leukocytes showed that GDP-fucose:N-acetyl-
glucosaminide 3-fucosyltransferase activity was demons-
trable in neutrophils, monocytes and lymphocytes but not
in platelets, whereas α-2-fucosyltransferase activity
was present in platelets but not in leukocytes (76;77).
    Several reports have appeared relating to abnormal-
ities of serum fucosyltransferase levels in leukaemia
(78,79,80). GDP-fucose: β-D-galactoside α-2-fucosyl-
transferase activity is high in serum from CML patients
in blast transformation and is frequently low in the
serum from AML patients. However, in both these conditions the enzyme level in serum correlates directly with
the platelet count in peripheral blood; hence the
depressed or elevated levels of α-2-fucosyltransferase
appear to arise from alterations in platelet number and
not from changes in the cellular expression of the enzyme
resulting directly from the leukaemic process (81).

α-3-Fucosyltransferase levels in serum are often markedly elevated in both CML and AML (80,81) and, in contrast to the 2-fucosyltransferase, this activity is frequently much higher in leukaemic blasts from AML patients than in normal neutrophils. The highest serum levels also occur in those patients with the largest numbers of circulating blasts (81), suggesting that, in leukaemia, at least part of the serum enzyme may be derived from the blasts.

Both X and sialyl-X determinants are strongly expressed on mature neutrophils but they are absent, or only weakly expressed, on leukaemic myeloblasts (82). In order to determine how the appearance of these antigenic determinants relates to the expression of the fucosyl- and sialyl- transferases Skacel et al.(75) examined normal neutrophils and leukaemic blasts from AML patients for these enzyme activities. Although when non-sialylated substrates were used much higher levels of 3-fucosyl-transferase activity were found in blasts than in mature neutrophils, with sialylated substrates the enzyme levels in blasts were not dissimilar from those in normal neut-rophils (Table III); the enzyme in the leukaemic cells from AML patients therefore resembles the enzyme purified from CML cells (69) in showing a marked preference for non-sialylated acceptors.

Table III. Fucosyl- and Sialyl-transferase Activities in Normal Neutrophils and Leukaemic Blasts from AML Patients

Cell Type	[14]Fucose* incorporated into:	
	Galβ1-4GlcNAc	NeuAcα2-3Galβ1-4GlcNAc
Neutrophils (n=7)	950	210
AML blasts (n=4)	7331	347

* c.p.m. incorporated per 10^6 cells.
SOURCE: Adapted from Skacel et al.(75).

Examination of sialyltransferase activities with the low-molecular-weight Type 2 substrate, Gal-β-1-4GlcNAc, revealed that, although no CMP-neuraminic acid: galact-osyl β-1-4-N-acetylglucosaminide 6'-sialyltransferase act-ivity was detectable in normal neutrophils, this enzyme was strongly expressed in the leukaemic cells. A much lower level of CMP-neuraminic acid:galactosyl β 1-4-N-acetylglucosaminide 3'-sialyltransferase activity was present in both the leukaemic blasts and the normal cells (Table IV). These results suggested that the reduced

expression of X in myeloblasts might be related to the
presence of the strong 6'-sialyltransferase which would
compete with both the 3'-sialyl- and the 3-fucosyl-
tranferases for the precursor Type 2 chain substrates.

Table IV. α-3'- and α-6'-Sialyl-transferase activities in
Neutrophils and Leukaemic Blasts from AML Patients

Cell type	6'-Sialyl-transferase	3'-Sialyl-transferase
	$(c.p.m.[^{14}C]$sialic acid transferred)	
Neutrophils (n=3)	- (0)	3067 (100)
AML Blasts (n=4)	16000 (77)	4564 (23)

Figures in parenthesis indicate percentage of total
activity
SOURCE: Adapted from Skacel et al.(75).

Changes Occurring During Differentiation of HL60 Cells.
The promyelocytic leukaemia cell line HL60 expresses the
X antigen (83) and has an 3-fucosyltransferase with a
preference for non-sialylated substrates (84). In order
to test the extent to which the differences between
leukaemic myeloblasts and mature cells might reflect the
changes occurring during normal myeloid maturation, HL60
cells were induced to differentiate along the granulo-
cytic pathway with dimethyl sulphoxide (DMSO) (85).
Morphological evidence of maturation was present in 80-
90% of treated cells and by 6 or 7 days most had progressed to myelocyte and metamyelocyte stages (75). Initially, as expected, the HL60 cells expressed X activity,
but not sialyl-X; however on exposure to DMSO, sialyl-X
determinants appeared, as measured by immunofluorescence
with a monoclonal anti-sialyl-X reagent, until by the
fifth day a population (20-30%) of the cells were strongly positive (Figure 5). Appearance of sialyl-X antigen
thus correlated with maturation of the HL60 cells.
    In uninduced HL60 cells elevated levels of 3-fucosyl-
transferase and 6'-sialyltransferase were comparable with
those found in leukaemic blasts. These levels fell within
24 hours exposure to DMSO and by day 5 activities measured with N-acetyllactosamine dropped to less than 40% of
those measured in the uninduced control cells. 3'-Sialyl-
transferase activity remained un-changed at 24 h. and
fell only slightly by day 6 (Figure 6). In control experiments, little change was observed in the UDP-
galactose: N-acetylglucosamine β-4-galactosyltransferase
activity of the HL60 cells measured after 5 days exposure
to DMSO, and KG1 cells, which do not differentiate in
response to DMSO, did not show a change in 3-fucosyl- or

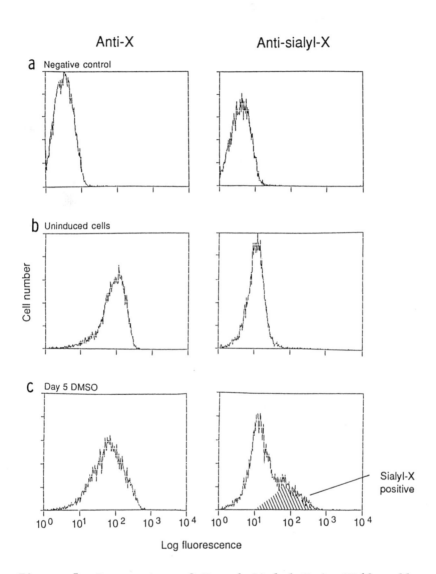

Figure 5. Expression of X and Sialyl-X in HL60 cells measured by cytofluorimetry. Negative controls **(a)**, uninduced HL60 cells **(b)** and cells induced to differentiate for 5 days with DMSO **(c)**. (Adapted from ref.75).

total (6'- plus 3'-)sialyltransferase activities after 48
h exposure to the DMSO reagent (75). The fall in the
sialyl- and fucosyl-transferase activities of the HL 60
cells could therefore be correlated with the differentiation process and not with an inhibitory effect of DMSO
on glycosyltransferase activity or expression.
    3-Fucosyltransferase activity in mature neutrophils
was similar when measured with fetuin either before or
after removal of sialic acid from the glycoprotein, with
only a marginal preference for the asialo-fetuin. In contrast, with HL60 cells the activity with asialo-fetuin
resembled that in leukaemic blasts in being approximately
5-10 times higher than with fetuin. Induction with DMSO
caused a marked fall in total activity measured with
asialo-fetuin and, as cell maturation progressed, there
was a change in the specificity of the 3-fucosyl-
transferase in the direction of a much less marked
preference for asialo-glycoprotein substrate (Figure 7).
    Examination of mononuclear cells isolated from
normal bone marrow revealed high levels of 6'-sialyl-
transferase and a 3-fucosyltransferase with similar acceptor substrate specificity to that measured in the
leukaemic cells (Skacel P.O.,unpublished data). The pattern of enzymes expressed in the leukaemic cells is,
therefore, most probably a reflection of the stage of
maturation arrest of the cells rather than the result of
a genetic, or epigenetic, change resulting directly from
the leukaemic process.

Inferences from the Studies on Haemopoietic Cells.
Oligosaccharide chains are built up by the concerted
action of many different glycosyltransferases and the
enzymes involved in the terminal glycosylation steps may
be influenced in their activity by the length, degree of
branching, internal structures and general conformation
of the chains, as well as by the nature of the carrier
glycoconjugate and other factors such as availability of
donor substrate, necessary co-factors etc. Explanations
for cell surface expression of carbohydrate antigens
based on enzyme results obtained with low- molecular
weight acceptors, or glycoprotein or glycolipid substrates that do not correspond to all the glycoconjugate
species likely to be present on the cells in question can
therefore only approximate to the conditions pertaining
in vivo. Nevertheless, the findings with normal leukocytes, leukaemic blasts and HL60 cells, suggest that in
myeloid cells competition resulting from changes in the
levels of glycosyltransferases acting on a common substrate, together with alterations in the fine specificity of
the enzymes, offer a plausible explanation for the
changes that occur in the expression of X and sialyl-X in
the course of cell maturation.
    In immature myeloid cells the 6'- and 3'-sialyl-
transferases and the 3-fucosyltransferase would each

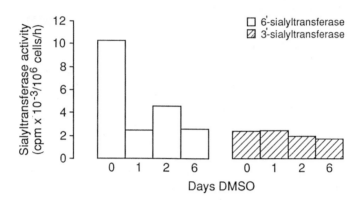

Figure 6. Sialyltransferase activities in HL60 cells induced to differentiate with DMSO. (Reproduced with permission from Ref. 75 Copyright 1991 The American Society of Haematology).

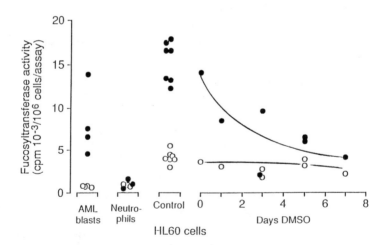

Figure 7. Fetuin (o) and asialofetuin (●) as substrates for the α-3-fucosyltransferases in AML blasts, normal neutrophils and HL60 cells induced to differentiate with DMSO. (Adapted from ref. 75).

compete for the available Type 2 chain end-groups but it can be postulated that the strongly expressed 6'-sialyl-transferase uses up much of the substrate at the expense of the other two enzymes. Moreover, since the fucosyl-transferase present in the immature cells does not uti-lise sialylated substrates, any NeuAc α 2-3Gal β 1-4-GlcNAc structures that are formed are not further converted to sialyl-X sequences (Figure 8). As the level of the 6'-sialyltransferase activity falls in the course of cell maturation the 3'-sialyltransferase would be able to com-pete more effectively for the Type 2 chains to provide a substrate for the the 3-fucosyltransferase species with the capacity to the transfer to 3'-sialyl-substituted acceptors and thereby give rise to the expression of sialyl-X on the cell surface (Figure 9). Consequently, in the early myeloid cells the 6'-sialyltransferase plays a regulatory role in the expression of X and sialyl-X even though it is not directly involved in the biosynthesis of these structures.

Evidence for modulation in cell surface carbohydrate expression resulting from alterations in the relative levels of fucosyl- and sialyl- transferases competing for the same substrate was earlier presented by Finne et al. (86) as the basis for the increased sensitivity of a mutant mouse melanoma clone to the fucose-binding lectin Lotus tetragonolobus.

Discussion and Conclusions

Much evidence has accumulated for the existence of a family of human GDP-fucose;galactosyl β1-4-N-acetylgluco-saminide α -3-fucosyltransferases with overlapping spec-ificities but distinct properties. These enzymes all **1)** transfer fucose to the O-3 position of N-acetylgluco-samine, **2)** fail to utilise glucose as a substrate to any appreciable extent and **3)** do not transfer fucose to the O-4 position of N-acetylglucosamine. These properties distinguish them from the blood-group Lewis gene-associ-ated 3/4-fucosyltransferase which has a much more limit-ed tissue distribution and, in addition to its 3-fucosyl transfer properties, transfers fucose to the O-4 position of N-acetylglucosamine to form the Le^a determinant.

Although the common property of the human 3-fucosyl-transferases is their capacity to synthesise X (Gal β -1-4[Fuc-α-1-3]GlcNAc) structures they can be differentiated on the basis of their capacity to use sialylated Type 2 substrates and hence to form sialyl-X (NeuAc α 2-3Gal β -1-4[Fuc-α-1-3]GlcNAc) structures. This difference was first observed between two GDP-fucose:galactosyl β 1-4-N-acetyl-glucosaminide   α -3-fucosyltransferases expressed   by Chinese hamster ovary cell (CHO) cell mutants (87). The two mutants expressed enzymes that differed in that one added fucose to the first internal N-acetylglucosamine residue in sialylated substrates and was sensitive to

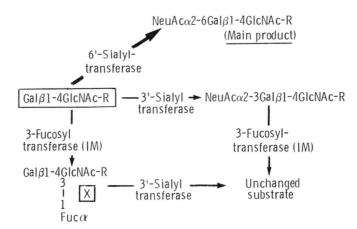

Figure 8. Biosynthetic pathways for the fucosylation and sialylation of Type 2 chain endings in immature myeloid cells. **IM,** immature myeloid form of α-3-fucosyltransferase.

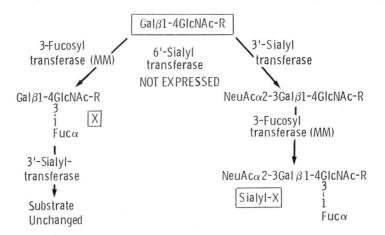

Figure 9. Biosynthetic pathways for the fucosylation and sialylation of Type 2 chain endings in mature myeloid cells. **MM,** mature myeloid form of α-3-fucosyltransferase

inhibition by N-ethylmaleimide whereas the other failed
to add fucose to the first N-acetylglucosamine residue in
sialylated substrates and was resistant to N-ethylmaleimide. The human 3-fucosyltransferases purified from
plasma, liver, and milk resemble each other, and one of
the CHO enzymes, with regard to their capacity to transfer to 3'-sialylated substrates (Table II) and their
sensitivity to sulphydryl reagents (55). The enzyme present in mature myeloid cells is similar to this group of
enzymes in that it utilises sialylated acceptors, although not as readily as the plasma-type 3-fucosyltrans-
ferases (Table II), but it differs in that it is relatively resistant to sulphydryl reagents (Skacel,P.O. unpublished data) and is expressed normally in the rare
individuals who have a deficiency of the 3-fucosyltrans-
ferase in plasma (64;88).
      The second species of 3-fucosyltransferase detected
in the immature myeloid cells failed to transfer fucose
to 3'-sialylated acceptors but resembled the "mature"
enzyme in its resistance to sulphydryl reagents (Skacel
P.O.,unpublished data). Moreover, neither the early or
late myeloid species reacted with antibodies that combined with the plasma-type enzymes (Johnson P.H., unpublished data). On this basis three different patterns
of activity are discernible; one found in plasma, milk
and liver (α-3-fucosyltransferase 1), a second found in
mature myeloid cells (α-3-fucosyl-transferase 2) and a
third found in immature myeloid cells (α-3-fucosyl-
transferase 3). These three specificity patterns are distinct from, and the tissue distribution of the enzymes is
different from, that of the Lewis-gene encoded 3/4-fuco-
syltransferase. Mollicone et al.,(61) examined a variety
of human tissue homogenates and body fluids for α-3-
fucosyltransferase activity with synthetic Type 1 and
Type 2 blood group H and sialylated acceptors and were
able to discern, in addition to the Lewis 3/4 pattern,
two patterns which they classifed as the myeloid pattern,
and the plasma-like pattern. The myeloid enzyme in normal
mature neutrophils described by these authors had very
low activity with sialylated acceptors and corresponds
more closely to the species found in our laboratory in
the immature cells (69;75); the reason for this discrepancy is not clear. However, the virtual failure of expression of sialyl-X by both the CML and AML cells
explains their lack of normal function in relation to
adhesion reactions requiring recognition of this ligand.
      GDP-fucose:galactosyl β1-4-N-acetylglucosaminide α-
3-fucosyltranferases with specificities similar to the
early myeloid form have been reported in other developing
tissues and in tumour tissues. The 3-fucosyltransferase
in the mouse teratocarcinoma stem cell F9 acts on asialo-
fetuin but not on fetuin; this enzyme disappears when the
stem cells are induced to differentiate into parietal
endoderm cells (89). Expression of X determinants, but

58   CARBOHYDRATE ANTIGENS

not sialyl-X determinants, therefore appears to be important in the early stages of development in the mouse.
The 3-fucosyltransferase purified from human neuroblastoma cells (90) similarly fails to transfer fucose to
substrates containing sialic acid linked either α-3- or
α-6- to a terminal β-galactosyl residue; therefore in
these cells sialyl-X would not be expressed on the cell
surface. However, inability to utilise sialylated
acceptors cannot be taken as a general property of 3-
fucosyltransferases expressed in tumour tissues because
accumulation of sialyl-X determinants is a characteristic
of many carcinomas (25;29;30;91).
     Whether the two different GDP-fucose:galactosyl-β-1-
4-N-acetylglucosamine α-3-fucosyltransferase species detected in myeloid cells arise from the expression of two
different genes or from some post-transcriptional or
post-translational modification of a single gene-encoded
protein remains to be established. Early studies on β-4-
galactosyltransferase (lactose synthetase) demonstrated
the change in acceptor specificity from N-acetyl-D-
glucosamine to D-glucose brought about by interaction
with a tissue specific protein encoded by a gene unrelated to the galactosyltransferase gene (92). More
recently studies on the gene encoding CMP-sialic acid: β-
galactoside α-2,6 sialyltransferase by Paulson and his
colleagues (93-94) have revealed the striking complexity
of this gene and its remarkably different expression in
different tissues; the gene contains at least three
promotors and produces at least five variant transcripts
in a tissue specific fashion. Joziasse et al.(95) have
also found that alternative splicing of the single gene
encoding murine α-3-galactosyltransferase results in four
transcripts that encode proteins that have the same
catalytic domain but differ in the lengths of their stem
region. Evidence for a regulator gene, independent of the
ABO locus, that controls the expression of the genes
encoding the blood-group associated glycosyltransferases
has also recently been presented (96).
     Whether the different sized mRNAs encoding the sialyl
transferase express enzymes with identical fine specificity requirements with regard to structurally related acceptor substrates has not been reported but it is con-
ceivable that a similar complexity of the fucosyl transferase genes might give rise to different sized transcripts that encode enzyme proteins sufficiently altered
in conformation and flexibility to influence their ability to combine with closely related, but differentially
substituted, oligosaccharide substrates.
     The numbers of different human genes required to
explain the apparently complex interrelationships of the
Lewis 3/4- and the other 3-fucosyl-transferases is still
not clear but the isolation of DNA encoding some of the
enzyme activities (8;44;97) is beginning to throw light
on this problem. The human DNA cloned by Kukowska et al.

(8) that encodes the Lewis blood group α-3/4-fucosyl-
tranferase was found to correspond to DNA sequences on
chromosome 19 where the Lewis locus had been mapped by
linkage studies (70). Cross hybridisation of this Lewis
cDNA with human genomic DNA from peripheral blood leukocytes (44) led to a different gene being isolated. The
"new" gene, which showed considerable structural homology
with the Lewis DNA, encoded an α-3-fucosyltransferase
that utilised Type 2, but not Type 1 substrates, and
could synthesise X but not sialyl-X determinants; this
enzyme on transfection into CHO cells did not give rise
to structures that bound to the adhesion molecule LEC-CAM
2 (ELAM 1). The properties of this enzyme resembled those
of the 3-fucosyltransferase in immature myeloid cells and
the gene, when used to probe mRNA from HL60 cells,
detected four distinct transcripts (44) suggesting the
possiblity that the fucosyltransferase(s) expressed by
HL60 cells might be encoded by the cloned DNA. Goetz et
al.(97) on the other hand, used direct-expression cloning
methods to isolate a gene (ELFT) that directed the expression of the LEC-CAM 2 (ELAM 1) ligand when transfected into previously non-binding cell lines; this gene
encoded a 3-fucosyltransferase that also utilised Type 2
and not Type 1 structures but, as judged from the expression of the LEC-CAM 2 ligand, differed in that it had
the capacity to transfer fucose to sialylated acceptors.
    Whether the three genes, the Le-gene encoded GDP-
fucose;N-acetyl-D-glucosaminide α-3/4-fucosyltransferase
and the two GDP-fucose:N-acetyl-D-glucosaminide α-3-fuc-
osyltransferases genes, together with tissue specific
processing of the gene products, can account for all the
manifestations of human 3-fucosyltransferase activity, in
embryonic, normal mature and tumour tissues, remains to
be established. However, the availability of these DNA
probes will provide tools for the isolation of possible
related genes, and for analysis of the promotors and en-
hancers involved in their expression. This information,
combined with in situ hybridisation to determine the
chromosomal and cellular localisation of the genes and
transfection and/or transgenic experiments to monitor
their function, should eventually allow a more complete
understanding of the complexities of the enzymes, and of
the precise roles of the biologically important cell
surface structures that they synthesise.

Acknowledgments

The unpublished results described in this article were
obtained in the Division of Immunochemical Genetics, MRC
Clinical Research Centre, Harrow, Middlesex, U.K. before
the closure of the Division in 1989. The support for this
work by the Medical Research Council and Leukaemia
research Fund (U.K.) is gratefully acknowledged.

Literature Cited

1. Kohler,G.; Milstein,C. Nature, 1975, 256, 495-497.
2. Hakomori,S. Ann.Rev.Immunol. 1984, 2, 103-126.
3. Feizi,T. Nature, 1985, 314, 53-57.
4. Watkins,W.M. Adv. Hum. Genet. 1980, 10, 1-136.
5. Beyer,T.A.; Sadler,J.E.; Rearick,J.I.; Paulson,J.C.; Hill,R.L. Adv.Enzymol. 1981, 52, 24-175.
6. Paulson,J.C.; Colley,K.J. J.Biol.Chem. 1989, 264, 17615-17618.
7. Yamamoto,F.; Clausen,H.; White,T.; Marken,J.; Hakomori,S. Nature 1990, 345, 229-233.
8. Kukowska-Latallo,J.F.; Larsen,R.D.; Nair,R.P.; Lowe,J.B. Genes & Dev. 1990, 4, 1288-1303.
9. Watkins,W.M.; Morgan,W.T.J. Nature 1957, 180, 1038-1040.
10. Rege,V.P.; Painter,T.J.; Watkins,W.M.; Morgan,W.T.J. Nature 1964, 203, 360-363.
11. Marr,A.M.S.; Donald,A.S.R.; Watkins,W.M.; Morgan,W.T.J. Nature 1967, 215, 1345-1349.
12. Longmore,G.D.; Schachter,H. Carbohydr.Res. 1982, 100, 365-392.
13. Lloyd,K.O.; Kabat,E.A.; Layug,E.J.; Gruezo,F. Biochemistry 1966, 5, 1489-1501.
14. Marr,A.M.S.; Donald,A.S.R.; Morgan,W.T.J. Biochem.J. 1968, 110, 789-791.
15. Rege,V.P.; Painter,T.J.; Watkins,W.M.; Morgan,W.T.J. Nature 1963, 200, 532-534.
16. Yang,H.; Hakomori,S. J.Biol.Chem. 1971, 246, 1192-1200.
17. Kobata,A. Methods Enzymol. 1972, 28, 262-271.
18. Lundblad,A. Methods Enzymol. 1978, 50, 226-235.
19. Holmes,E.H.;Ostrander,G.K.; Hakomori,S. J.Biol.Chem. 1985,7619-7627.
20. Gunson,H.H.; Latham,V. Vox Sang. 1972, 22, 344-353.
21. Huang,L.C.; Civin,C.K.; Magnani,J.L.; Shaper,J.H.; Ginsburg,V. Blood 1983, 61, 1020-1023.
22. Solter,D.; Knowles,B.B. Proc.Natl.Acad.Sci.U.S.A. 1978, 75, 5565-5569.
23. Rauvala,H. J.Biol.Chem. 1976, 251, 7517-7520.
24. Hanisch,F.; Uhlenbruck,G.; Dienst,C. Eur.J.Biochem. 1984, 144, 467-474.
25. Fukushima,K.; Hirota,M.; Terasaki,P.; Wakishaka,A.; Togashi,H.; Chia,D.; Suyama,N.; Fukushi,Y.; Nudelman,E.; Hakomori,S. Cancer Res. 1984, 44, 5279-5285.
26. Fukuda,M. Biochim.Biophys.Acta 1985, 780, 119-150.
27. Holmes,E.H.; Ostrander,G.K.; Hakomori,S. J.Biol.Chem. 1986, 261, 3737-3743.
28. Macher,B.A.; Buchler,J.; Scudder,P.; Knapp,W.; Feizi,T. J.Biol.Chem. 1988, 263, 10186-10191.
29. Hakomori,S. Cancer Res. 1985, 45, 2405-2414.
30. Fukushi,Y.; Nudelman, E.; Levery,S.; Hakomori,S.; Rauvala,H. J.Biol.Chem. 1984, 259, 10511-10517.

31.Matsushita,Y; Cleary,K.R.; Ota,D.M.; Hoff,S.D.;
   Irimura,T.Lab. Invest. **1990**, 63, 780-791.
32.Civin,C.L.; Mirro,J.; Banquerigo,M.L. Blood **1981**, 57,
   842-845.
33.Strauss,L.C.; Stuart,R.K.; Civin,C.L. Blood **1983**, 61,
   1222-1231.
34.Gooi,H.C.; Feizi,T.; Kapadia,A.; Knowles,B.B.;
   Solter,D.; Evans,M.J. Nature **1981**, 292, 156-158.
35.Stoolman,L.M. Cell **1989**, 56, 907-910.
36.Osborn,L. Cell **1990**, 62 3-6.
37.Springer,T.A.; Lasky,L.A. Nature **1991**, 349, 196-197.
38.Brandley,B.K.; Swiedler,S.J.; Robbins,P.W. Cell **1990**,
   63, 861-863.
39.True,D.D.; Singer,M.S.; Lasky,L.A.; Rosen,S.D. J.Cell
   Biol.**1990**, 111, 2757-2764.
40.Lowe,J.B.; Stoolman,L.M.; Nair,R.P.; Larsen,R.D.;
   Berhend,T.L.; Marks,R.M. Cell **1990**, 63, 475-484,
41.Phillips,M.L.; Nudelman,E.; Gaeta,F.C.; Perez,M.;
   Singhai,A.K.; Hakomori,S.; Paulson,J.C. Science, **1990**,
   250, 1130-1132.
42.Waltz,G.; Aruppo,A.; Kolanus,W.; Bevilacqua,M.;
   Seed,B. Science **1990**, 250, 1132-1135.
43.Tiemeyer,M.; Swiedler,S.J.; Ishihara,M.; Moreland,M.;
   Schweingruber,H.; Hirtzer,P.; Brandley,B.K. Proc.
   Natl.Acad.Sci,U.S.A. **1991**, 88, 1138-1142.
44.Lowe,J.B.; Kukowska-Latallo,J.F.; Nair R.P.;
   Larsen,R.D.;Marks,R.M.; Macher,B.A.; Kelly R.J.;
   Ernst,L.K. J.Biol.Chem. **1991**, 266, 17467-17477.
45.Larsen,E.; Palabrica,T.; Sajer,S.; Gilbert,G.E.;
   Wagner,D.D.; Furie,B. Cell **1990**, 63, 467-474.
46.Winkelhake,J.L. Glycoconjugate J. **1991**, 8, 381-386.
47.Shen,L.; Grollman,E.F.; Ginsburg,V.
   Proc.Natl.Acad.Sci.U.S.A. **1968**, 59, 224-230.
48.Chester M.A.; Watkins W.M. Biochem.Biophys.Res.Commun.
   1969, 34, 835-842.
49.Schenkel-Brunner,H.; Chester,M.A.; Watkins,W.M.
   Eur.J.Biochem. **1972**, 30, 269-277.
50.Race,R.R.; Sanger,R. Blood Groups in Man.
   Blackwell, Oxford, U.K. 5th edn. **1968**; pp 305-330.
51.Johnson,P.H.; Yates,A.D.; Watkins W.M.
   Biochem.Biophys.Res.Commun. **1981**, 100, 1611-1618.
52.Prieels J-P.; Monnom,D.; Dolmans,M.; Beyer,T.A.;
   Hill,R.L. J.Biol.Chem. **1981**, 256, 10456-463.
53.Johnson,P.H.; Watkins,W.M.; Donald,A.S.R. In Proc.9th
   Int.Symp.Glycoconjugates; Editors, Montreuil,J.;
   Verbert,A.; Spik,G.; Fournet B.; Lille, France, **1989**,
   Abstract E 107.
54.Watkins W.M.; Greenwell,P.; Yates,A.D.; Johnson,P.H.
   Biochimie **1988**, 70, 1597-1611.
55.Johnson,P.H. Ph.D.Thesis, University of London **1988**,
56.Paulson,J.C.;Prieels,J-P; Glasgow,L.R.; Hill,R.L.
   J.Biol.Chem. **1978**, 253, 5617-5624.

57.Johnson,P.H.; Watkins,W.M. Biochem.Soc.Trans. 1985, 13,1119-1120
58.Johnson,P.H.; Watkins,W.M. Biochem.Soc.Trans. 1982, 10, 445-446.
59.Johnson,P.H.; Watkins,W.M. In Proc.10th Int.Symp. Glycoconjugates; Editors; Sharon,N.; Lis,H.; Duksin,D.; Kahane,I. Symposium Organising Committee, Jerusalem, Israel, 1989, Abstract 147,pp 214-215.
60.Hanisch,F-G.; Misakos,A.;Schroten,H.; Uhlenbruck,G. Carbohydr.Res. 1988, 178, 23-28.
61.Mollicone,R.; Gibaud,A.; Francis,A.; Ratcliffe,M.; Oriol,R. Eur.J.Biochem. 1990, 191, 169-176.
62.Grollman,E.F.; Kobata,A; Ginsburg,V. J.Clin.Invest 1969, 48, 1489-1494.
63.Greenwell,P.; Johnson,P.H.; Edwards,J.M.; Reed,R.M.: Moores,P.P.; Bird,A.; Graham,H.A.; Watkins,W.M. Blood Transf.Immunohaematol. 1986, 29, 233-249.
64.Caillard,T.; Le Pendu,J.; Ventura,M.; Mada,M.; Rault,G.; Mannoni,P.; Oriol,R. Expl.Clin.Immunogenet. 1988, 5, 15-23.
65.Oriol,R; Mollicone,R.; Masri,R.; Whan,I.; Lovric,V.A. Glycoconjugate J. 1991, 8, Abstract 4.7, p 147.
66.Spooncer,E.; Fukuda,M.; Klock,J.C.; Oates,J.E.; Dell,A. J.Biol.Chem. 1984, 259, 4792-4801.
67.Fukuda,M.; Spooncer,E.; Oates,J.E.; Dell,A.; Klock,J.C. J.Biol.Chem. 1984, 259, 10925-10935.
68.Fukuda,M.N.: Dell,A.; Oates,J.E.; Wu P.; Klock,J.C.; Fukuda,M. J.Biol.Chem. 1985, 260, 1067-1082.
69.Johnson,P.H.; Watkins,W.M. Biochem.Soc.Trans. 1987, 15,396.
70.Weitkamp,L.R.; Johnston,E.: Guttertormsen,S.A. Cytogenet.Cell Genet. 1974, 13, 174-184.
71.Guerts van Kessel,A.; Tetteroo,P.; van Agthoven,T.; Paulussen,R.; van Dongern,J.; Hagemeijer,A.; Von dem Borne,A. J.Immunol. 1984, 133, 1265-1269.
72.Tetteroo,P.A.T.; de Heij,H.T.; Van den Eijnden,D.H.; Viser,F.J.; Schoenmaker,E.; Geurts van Kessel,A.H.M. J.Biol.Chem. 1987, 262, 15984-15989.
73.Coulin,P.; Mollicone,R.; Grisard,M.C.; Gibaud,A.; Ravise,N.; Feingold,J.; Oriol,R. Cytogenet.Cell Genet. 1991 56, 108-111.
74.Chou,T.H.; Murphy,C.; Kessel,D. Biochem.Biophys.Res. Commun. 1977, 74, 1001-1006.
75.Skacel,P.O.; Edwards,A.J.; Harrison,C.T.; Watkins,W.M. Blood 1991, 78, 1452-1460.
76.Skacel,P.O.; Watkins,W.M. Glycoconjugate J. 1987,4, 267-272.
77.Mollicone,R.; Caillard,T.; Le Pendu J.; Francois,A.; Sansonette,N.; Villarroya,H.; Oriol,R. Blood 1988, 71 1113-1119.
78.Khilanani,P.; Chou,T-H,; Lomen,P.L.; Kessel,D. Cancer Res. 1977, 37 2557-2559.
79.Kuhns,W.J; Oliver,R.T.D.; Watkins,W.M.; Greenwell,P. Cancer Res. 1980, 40, 268-275.

80.Kessel,D.; Ratanatharathorn,V.; Chou T-H. Cancer Res.
   1979, 39, 3377-3380.
81.Skacel,P.O.; Watkins,W.M. Biochem.Soc.Trans. **1988**, 16,
   1034-1035.
82.Tabilio,A.; Del Canizo,M.C.; Henri,A.; Guichard,J.;
   Mannoni,P.; Civin,C.I.; Testa,U.; Rochant,H.;
   Vainchenker,W.; Breton-Gorius,J. Br.J.Haematol. **1984**,
   58, 697-710.
83.Symington,F.W.; Hedges,D.L.; Hakomori,S. J.Immunol.
   1985, 134, 2498-2506.
84.Potvin,B.; Kumer,R.; Howard,D.R.; Stanley,P.
   J.Biol.Chem. **1990**, 256, 1615-1622.
85.Collins,S.J.; Ruscetti,F.W.; Gallanger,R.E.;
   Gallo,R.C. Proc.Natl.Acad.Sci. U.S.A. **1979**, 75, 2458-
   2462.
86.Finne,J.; Burger,M.M.; Prieels,J-P. Cell Biol. **1982**,
   92 277-282.
87.Campbell,C.; Stanley,P. J.Biol.Chem. **1984**, 259, 11208-
   11214.
88.Johnson,P.H.; Skacel,P.O.; Greenwell,P.; Watkins,W.M.
   Biochem.Soc.Trans. **1989**, 17, 133-134.
89.Muramatsu,H.; Muramatsu,T. FEBS Lett. **1983**, 163, 181-
   184.
90.Foster,C.S.; Gillies,D.R.B.; Glick,M.G. J.Biol.Chem.
   1991, 266, 3526-3531.
91.Magnani,J.L.; Nilsson,B.; Brockhaus,M.; Zopf,D.;
   Steplewski,Z.; Koprowski,H.; Ginsburg,V. J.Biol.Chem.
   1982, 14,365-369.
92.Brodbeck,U.; Denton,W.L.; Tanahashi,N.; Ebner,K.E.
   J.Biol.Chem. **1967**, 242, 1391-1397.
93.Paulson,J.C.; Weinstein,J.: Schauer,A. J.Biol.Chem.
   1989, 264 10931-10934.
94.Wen,D.X.; Svensson,E.C.; Paulson,J.C. Glycoconjugate
   J. **1991**, 8, Abstract 4.3, p.146.
95.Joziasse,D.H.; Shaper,J.H.; Jun,D.; N.L.Shaper
   Glycoconjugate J. **1991**, 8 Abstract 4.18, p.150.
96.Dorscheid,D.; Friedlander,P.; Price,G. Glycoconjugate
   J. **1991**, 8, Abstract 4.19, p.151.
97.Goetz,S.E.; Hession,C.; Goff,D.; Griffiths,;
   Tizard,R.; Newman,B.; Chi-Rossi,G.; Lobb,R. Cell **1991**,
   63, 1349-1356.

RECEIVED March 25, 1992

Chapter 5

Vibrio cholerae Polysaccharide Studies

Bengt Lindberg

Department of Organic Chemistry, Arrhenius Laboratory, Stockholm
University, S–10691 Stockholm, Sweden

Structural studies on the O-antigens from *Vibrio cholerae* O:1, O:2, O:3
and O:21 will be discussed. They are composed of mono- or
oligosaccharide repeating units and contain unusual sugar components,
namely, 3,6-dideoxy-L-*arabino*-hexose, D-*glycero*-D-*manno*-heptose, 4-
amino-4,6-dideoxy-D-mannose, 2,4-diamino-2,4,6-trideoxy-D-glucose,
and 5,7-diamino-3,5,7,9-tetradeoxy-L-*glycero*-L-*manno*-nonulosonic acid.
They also contain some unique or unusual substituents, namely,
acetamidino groups and N-linked 3-deoxy-L-*glycero*-tetronyl and 3,5-
dihydroxyhexanoyl groups.

The species *Vibrio cholerae* is divided into several serogroups on the basis of their O-
antigens. In addition to serogroup O:1, which causes Asian cholera, there are some 80
other serogroups (*1*), which cause similar but less severe diseases. A
lipopolysaccharide (LPS) from a gram-negative bacterium is composed of three parts
(*1*). The lipid part, lipid A, which is the endotoxically active part, is linked *via* a core
oligosaccharide to the O-specific polysaccharide. The lipid A is quite similar in most
gram-negative bacteria, and that from *V. c.* O:1 (*2*) does not differ much from those
found in Enterobacteriaceae.

O-Polysaccharide-Core-Lipid A

1

The core part is less conservative, and about a dozen related structures have
been found in Enterobacteriaceae. In all cores there is a 3-deoxy-D-*manno*-octulosonic
acid (Kdo) residue, linking the core to lipid A. This sugar is labile in acidic solution and
was for a long time overlooked in the *V. c.* O:1 LPS. Its presence was, however,
established by Brade *et al.* (*3*). The structure of the core in *V. c.* O:1 is not known, but
it contains Kdo, D-glucosamine (with a free amino group), D-fructose, D-glucose, and
L-*glycero*-D-*manno*-heptose. A disaccharide (2) is obtained on partial hydrolysis with
acid (*4*). The O-antigen, which is composed of mono- or oligosaccharide repeating
units, determines the O-antigenic specificity of the bacterium, and there is an enormous
structural variation.

0097–6156/93/0519–0064$06.00/0

$$\alpha\text{-}\mathrm{D\text{-}Glc}p\mathrm{N\text{-}(1{\rightarrow}7)\text{-}L,D\text{-}Hep}$$
2

In the following, structural studies on four O-polysaccharides from *Vibrio cholerae* will be briefly discussed. The structures have little in common, except that they all contain unusual sugar components.

Vibrio cholerae O:1 O-Antigen

Because of the problems caused by Asian cholera, the LPS of *V.c.* O:1 has been extensively studied, but it was not until 1979 that the correct structure of its O-polysaccharide was published (5). The ^{13}C n.m.r. spectrum of the O-antigen showed 10 signals, indicating a simple, regular structure. Six of these were assigned to a carboxyl group (δ 179.1), an anomeric carbon (δ 102.7), a hydroxymethyl group (δ 60.1), a carbon linked to nitrogen (δ 55.1), a methylene group (δ 38.1), and a C-methyl group (δ 18.8). Hydrolysis with acid yielded a polymer of an amino-6-deoxyhexose and a C_4 carboxylic acid.The acid was identified as 3-deoxy-L-*glycero*-tetronic acid from its ^{13}C and ^{1}H n.m.r. spectra and optical rotation.On solvolysis of the original polysaccharide with liquid hydrogen fluoride, when glycosidic linkages are cleaved but amide linkages are intact, 4-amino-4,6-dideoxy-D-*manno*-hexose, *N*-acylated with this acid, was obtained and identified, using n.m.r. and g.l.c.-m.s. of its alditol acetate. Methylation analysis, also with hydrolysis in liquid hydrogen fluoride,showed that the sugar was linked through *O*-2 in the polysaccharide, and n.m.r. that it was α-linked. The O-polysaccharide is consequently composed of monosaccharide repeating units with the structure 3. Even if this structure is simple, earlier studies were complicated by the fact that most of the 4-amino sugar was decomposed during hydrolysis with acid and was assumed to be a minor component.On solvolysis with liquid hydrogen fluoride, however, the *N*-acylated sugar obtained is quite stable.

3

On deamination of the *N*-deacylated polysaccharide the amino sugar residues were transformed into a mixture of D-rhamnopyranosyl and L-allofuranosyl residues by opening of an intermediate epoxonium ion either at C-4 or at C-5 (Scheme 1), as demonstrated by methylation analysis of the polymeric product.

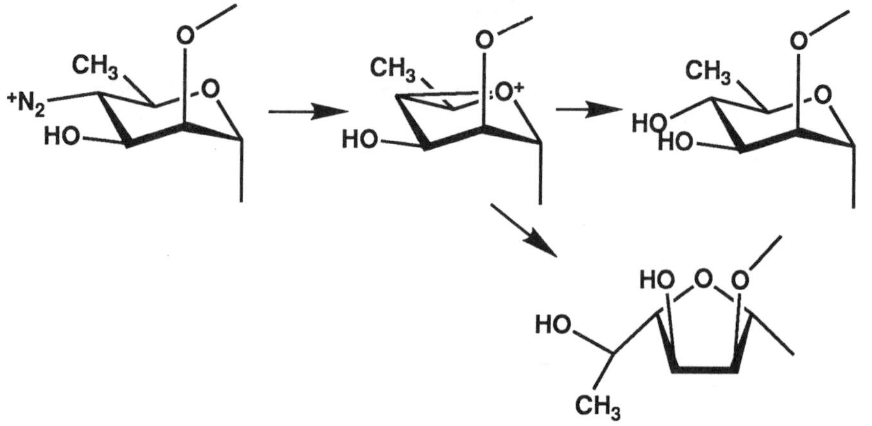

Scheme 1. Deamination of the *N*-deacylated *V. c.* O:1 O-polysaccharide.

V. c. O:1 occurs as two immunologically distinct strains, Ogawa and Inaba, which seem to contain the same O-polysaccharide. The nature of the LPS determinants specific for these strains is, however, still obscure. It has been found(*6*) that the gene clusters which express the biosyntheses of the O-antigens in Inaba and Ogawa contain 15 and 15+5 kilobases, respectively, indicating that the Ogawa O-antigen contains some structural feature that is lacking in the Inaba O-antigen.

Vibrio cholerae O:2 O-Antigen

The *V. c.* O:2 O-polysaccharide (*7*) on hydrolysis yielded 2-acetamido-2,6-dideoxy-D-glucose (*N*-acetyl-D-quinovosamine) and D-galactose in the ratio 1:1. Methylation analysis and ^{13}C and ^{1}H n.m.r. spectra demonstrated that these sugars were β-pyranosidic and linked through *O*-4, and also indicated that the polysaccharide contained a keto sugar with one carboxyl group, one acetamidino group [-NH-$\underline{C}$=N($\underline{C}$H$_3$), δ 168.41 and 19.64], one acetamido group, one methyl and one methylene group. Two 5,7-diamino-3,5,7,9-tetradeoxy-nonulosonic acids had previously been found by Kochetkov´s group (*8*), and our sugar most probably belonged to the same class.

The acetamidino group was transformed into an acetamido group by treatment with aqueous triethyl amine. Methanolysis of the thus modified polysaccharide yielded a methyl ester methyl glycoside of a disaccharide, composed of the unknown sugar and D-galactose. Smith degradation of this disaccharide yielded **4**, and a related

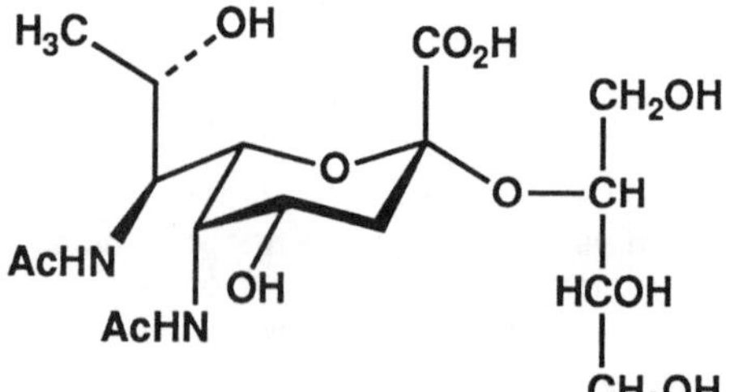

4

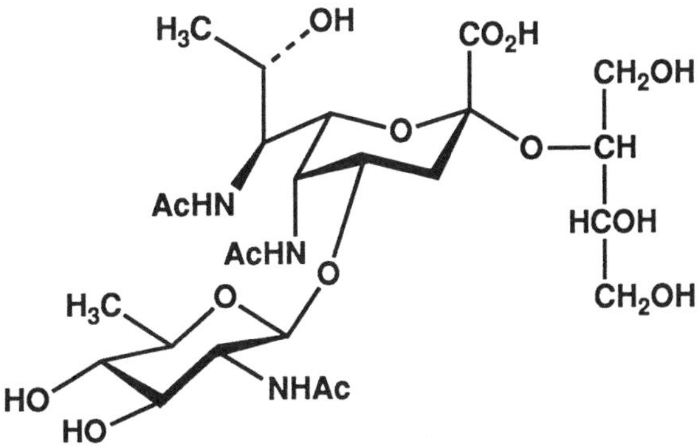

5

trisaccharide (**5**) was obtained on Smith degradation of the modified polysaccharide. These substances were characterized by 2-D n.m.r.and optical rotation, and the acid identified as 5,7-diacetamido-3,5,7,9-tetradeoxy-L-*glycero*-L-*manno*-nonulosonic acid. Comparison of the [13]C n.m.r. spectra of the original and the modified polysaccharide demonstrated that the acetamidino group is in the 5-position of the acidic sugar, which is 5-acetamidino-7-acetamido-3,5,7,9-tetradeoxy-L-*glycero*-L-*manno*-nonulosonic acid (**6**). The parent acidic sugar is thus one of the two members of this class previously found(*8*). From the combined results it was concluded that the O-polysaccharide is composed of trisaccharide repeating units with the structure **7**, in which Sug stands for **6**.

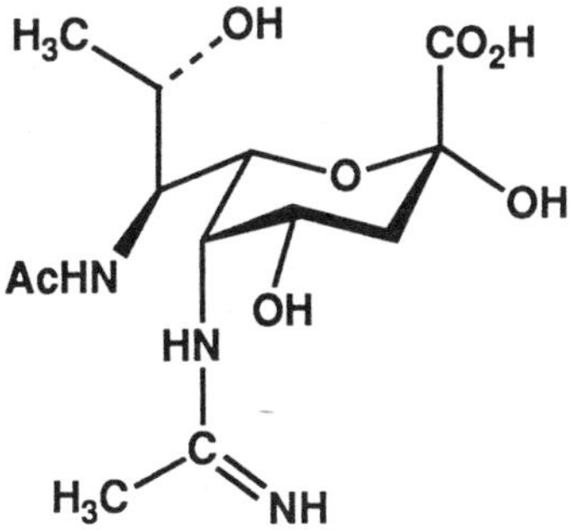

6

→4)-β-D-Qui*p*NAc-(1→4)-β-Sug*p*-(2→4)-β-D-Gal*p*-(1→
7

Vibrio cholerae O:3 O-Antigen

The O-polysaccharide from *V. c.* O:3 on hydrolysis yielded 3,6.dideoxy-L-*arabino*-hexose (ascarylose), D-*glycero*-D-*manno*-heptose, and 2-amino-2,6-dideoxy-L-galactose (L-fucosamine) (*9*). The ^{1}H and [13]C n.m.r. spectra demonstrated the presence of two *N*-acetyl groups and a fourth aldose component. In a COSY spectrum the presence of a spin system deriving from a 3,5-dihydroxyhexanoyl group (**8**) was detected. From the unassigned signals it was further concluded that the polysaccharide contained a 6-deoxyhexose with two amino groups. Methanolysis, followed by *N*-

acetylation yielded, *inter alia,* a methyl glycoside, identified by its n.m.r. spectra and optical rotation as methyl 2,4-diacetamido-2,4,6-trideoxy-α-D-glucopyranoside (**9**).

8   9

10

Methylation analysis revealed that the ascarylose was terminal, the *N*-acetyl-L-fucosamine was linked through *O*-4, and the heptose through *O*-2 and *O*-3. On treatment with acid under mild conditions ascarylose was hydrolyzed off, and methylation analysis of the thus modified polysaccharide demonstrated that it had been linked to *O*-3 of the heptose residue. Some low-molecular weight substances were also obtained in low yields during the hydrolysis. One of these, after reduction with sodium borodeuteride and methylation, on e.i.-m.s. gave the fragments indicated in formula **10**, demonstrating that the 3,5-dihydroxyhexanoyl residue is linked to *N*-4 of the diaminohexose.

Other products were identified as the disaccharide alditol **11** and the trisaccharide alditol **12**.

D-α-D-Hep*p*-(1→4)-L-FucNAc-ol-1d
11
β-Sug*p*-(1→2)-D-α-D-Hep*p*-(1→4)-L-FucNAc-ol-1d
12

The L-FucNAc and Asc were, according to n.m.r. evidence, both α–linked. The combined results therefore indicate that the *V. c.* O:3 O-polysaccharide is composed of tetrasaccharide repeating units with the structure **13**.

$\rightarrow$2)-D-α-D-Hep$p$-(1$\rightarrow$4)-α-L-Fuc$p$NAc-(1$\rightarrow$3)-β-Sug$p$-(1$\rightarrow$
3
$\uparrow$
1
$\alpha$$-Ascp$

13

In this structure it is assumed that the 2,4-diamino-2,4,6-trideoxy-β-D-glucopyranosyl residue (Sug) is linked through O-3, the only available position in the sugar moiety. There are, however, examples of polysaccharides containing amino sugars acylated by hydroxycarboxylic acids, in which the adjacent sugar is not linked to the amino sugar but to a hydroxyl group in the acyl group. In order to differentiate between these possibilities the methylated polysaccharide was subjected to f.a.b.-m.s. in the positive mode. It was known that methylated glycoconjugates of high-molecular weight, containing 2-acetamido-2-deoxyhexosyl residues, are cleaved under these conditions (*10*), giving A$_1$-type ions. The methylated polysaccharide gave ions of m/z 994 and 622, which are the expected A$_1$ ions (**14** and **15**) formed from the "non-reducing" end of the polysaccharide. The results support the sequence of sugars given above and indicate that **13** is actually the biological repeating unit.

CH$_3$

$$CH_3CH(OMe)CH_2CH(OMe)CH_2\cdot\overset{\overset{\textstyle O}{\|}}{C}\cdot\underset{Me}{N}$$

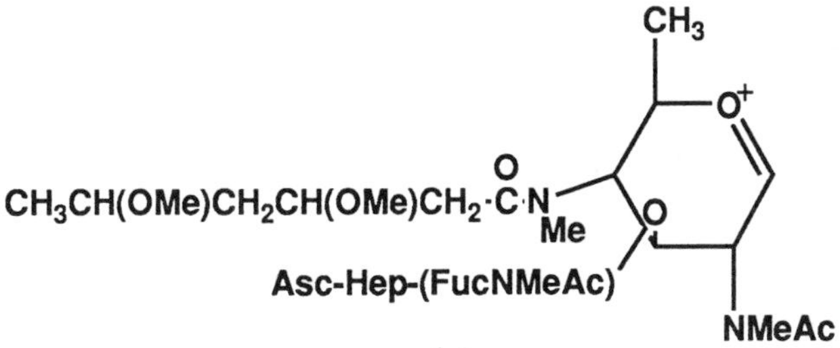

Asc-Hep-(FucNMeAc)

NMeAc

14

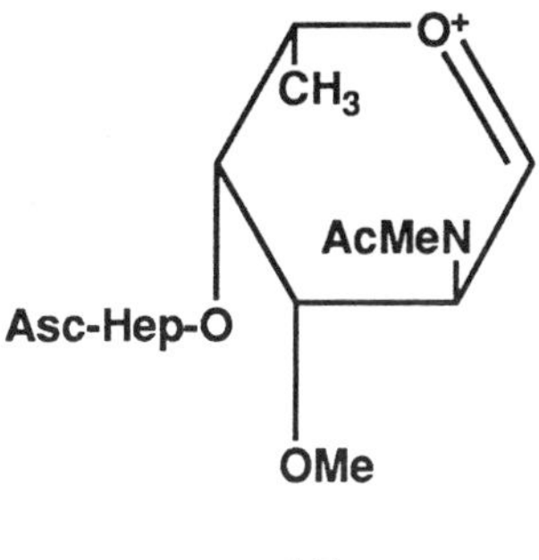

CH$_3$

AcMeN

Asc-Hep-O

OMe

15

CH₃CH(OMe)CH₂CH(OMe)CH₂C

16

Secondary fragments are generally weak on f.a.b.-m.s. of carbohydrates. One exception is the A_1-type fragments discussed above, which eliminate the substituent in the 3-position (*10*), either as methanol or as a methylated mono- or oligosaccharide. A strong fragment of *m/z* 355 probably has the structure **16**, and is formed from **14** by elimination of the trisaccharide Asc-Hep-FucNAc. The result thus indicates that the diaminohexose is actually linked through *O*-3, as in formula **13**.

Attempts to isolate the 3,5-dihydroxyhexanoic acid in order to determine its absolute configuration were not successful, most probably because of β-elimination. The (3*S*),(5*S*)-isomer of this acid is present in berries of *Sorbus aucuparia* (mountain ash) as the β-D-glucopyranoside of its δ-lactone (*11*). 2-*N*-Acetyl-4-*N*-[(*S*)-3-hydroxybutyryl]-2,4,6-trideoxy-D-glucose is a component of the O-polysaccharide from a serogroup of *Pseudomonas aeruginosa*. (*12*).

Vibrio cholerae O:21 O-Antigen

The *V. c.* O:21 O-polysaccharide (*13*) on hydrolysis yielded equimolecular amounts of L-rhamnose, *N*-acetylglucosamine, *N*-acetylgalactosamine, and D-*glycero*-D-*manno*-heptose. N.m.r. demonstrated that all sugars were pyranosidic, that both amino sugars were β-linked, and that one of the two sugars with *manno*-configuration was α-linked and the other β-linked.Methylation analysis revealed that L-Rha*p* and β-D-Gal*p*NAc were terminal, β-D-Glc*p*NAc was linked through O-3 and the heptose through the 3-, 4-, and 7-positions.Smith degradation gave a linear polysaccharide in which the heptose was linked through O-7, and n.m.r. proved it to be β-linked.The L-rhamnopyranosyl residue is consequently α-linked. On mild acid hydrolysis mild L-rhamnosyl groups were hydrolyzed off, and 2,3,6-tri-*O*-methylheptose appeared in the methylation analysis. From these results it was concluded that the polysaccharide is composed of tetrasaccharide repeating units with the structure **17**.

β-D-Gal*p*NAc

1

↓

4

→3)-β-D-Glc*p*NAc-(1→7)-D-β-D-Hep*p*-(1→

3

↑

1

α-L-Rha

17

ACKNOWLEDGMENTS

It is a pleasant duty to thank members of our research group, listed in the bibliography, for their contributions to practical and theoretical aspects of this work. The work has been supported by the Swedish Medical Research Council and the Swedish National Board for Technical Development.

LITERATURE CITED

1. Sakazaki, R.; Donovan, J. *Methods Microbiol.* **1984**, *16*, 271-289.
2. Broady, K. W.; Rietschel, E. Th.; Lüderitz, O. *Eur. J. Biochem.* **1981**, *115*, 463-468.
3. Brade, H.; Galanos, C.; Lüderitz, O. *Eur. J. Biochem.* **1983**, *131*, 195-200.
4. Kaca, W.; de Jongh-Leuvenink, J.; Zähringer, U.; Rietschel, E. Th.; Brade, H.; Verhoef, J.; Sinnwell, V. *Carbohydr. Res.* **1988**, *179*, 289-299.
5. Kenne, L; Lindberg, B.; Unger, P.; Holme, T; Holmgren, J. *Carbohydr. Res.* **1979**, *68*, C14-C16.
6. Manning, P. A.; Heuzenroeder, M. W.; Yeadon, J.; Leavesley, D. I.; Reeves, P. R.; Rowley, D. *Infect. Immun.* **1986**, *53*, 272-277.
7. Kenne, L.; Lindberg, B.; Schweda, E.; Gustafsson B.; Holme, T. *Carbohydr. Res.* **1988**, *180*, 285-294.
8. Knirel, Yu. A.; Kochetkov, N. K. *FEMS Microbiol. Rev.* **1987**, *46*, 381-385.
9. Chowdhury, T. A.; Jansson, P.-E.; Lindberg, B.; Lindberg, J.; Gustafsson, B.; Holme, T. *Carbohydr. Res.* **1991**, *215*, 303-314.
10. Dell, A. *Adv. Carbohydr. Chem. Biochem.* **1987**, *45*, 19-72.
11. Tschesche, R.; Hoppe, H-J.; Snatzke, G.; Wulff, G.; Fehlhaber, H.-W. *Chem. Ber.* **1971**, *104*, 1420-1428.
12. Knirel, Yu. A.; Vinogradov, E. V.; Shaskov, A. S.; Wilkinson, S. G.; Takara, Y.; Dmitriev, B. A.; Kochetkov, N. K.; Stanislavsky, E. S.; Mashilova, G.M. *Eur. J. Biochem.* **1986**, *155*, 659-669.
13. Ansari, A. A.; Kenne, L.; Lindberg, B.; Gustafsson, B.; Holme, T. *Carbohydr. Res.* **1986**, *150*, 213-219.

RECEIVED March 25, 1992

Chapter 6

Distinguishing Antibodies That Bind Internal Antigenic Determinants from Those That Bind the Antigenic Chain-Terminus Only

Cornelis P. J. Glaudemans, Larry G. Bennett, Apurba K. Bhattacharjee, Eugenia M. Nashed, and Thomas Ziegler

Section on Carbohydrates, National Institute of Diabetes and Digestion and Kidney Diseases, National Institutes of Health, Bethesda, MD 20892

The binding of monoclonal antibodies to polymeric antigens is discussed. Once the immuno-determinant, and its affinity for the antibody, is known, a measurement using the entire polymeric antigen and the Fab or Fab' fragment of the antibody can quickly establish if the antibody can read that determinant at any position in the antigen's polymeric chain, or if it is limited to binding that determinant only when it is located at the chain terminus.

Many (polymeric) antigens (particularly polysaccharides) contain repetitive groupings, certain sequences of which make up the immuno-determinant. Antibodies directed towards such a determinant may read (be able to bind to) the determinant when it occurs internally in the chain, or they may be able to do so only when it occurs at the chain termini of the antigen. This paper will offer a simple method to distinguish these two modes of binding. The illustration of the method will involve polysaccharide-monoclonal antibody systems, but the method is equally applicable to any antigen-antibody system.

POLYSACCHARIDE- AND ANTIBODY STRUCTURE
Polysaccharides, unlike proteins, are secondary gene products. For this reason, their molecular weight is not sharply defined, and they show a molecuar weight distribution, which is usually Gaussian. Due to the mechanism of their biosynthesis, hetero-polysaccharides (those polysaccharides consisting of different constituent sugars, and/or having different type linkages) show an overall chemical structure which consists of a precisely regular repeating substructure. Homo-polysaccharides (containing only one kind of carbohydrate, in a uniform linkage) also - of course - display a precisely repeating regularity in their structure.

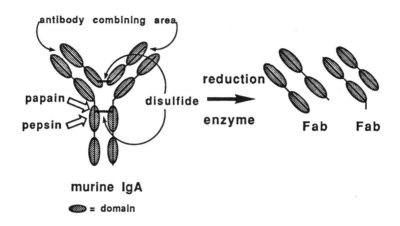

The size of the combining area in antibodies with anti-carbohydrate specificity varies somewhat, to accomodate of from four to six sugar residues (1-3). Antibodies are made up of two identical heavy (H) chains each of molecular weight *ca.* 50,000 Daltons and two identical light (L) chains of molecular weight *ca.* 25,000 Daltons each. The four chains each possess a number of globular domains, each domain containing approximately 100 amino acids. The H chain has four (five in IgM's), and the L chain has two such domains. The N-terminal domains of each H,L pair together harbor a combining area at the solvent-exposed end of the molecule. Thus, each H_2L_2 unit possesses two combining areas. The N-terminal domains of both the H and the L chain vary from immunoglobulin to immunoglobulin, while the remainder of both H and L chain are constant within a given class. That variability of the first N-terminal domain of both the H and the L chain results in the expression of a unique specificity of the immunoglobulin. This variability resides particularly in certain key areas of the variable domain, and those are called hypervariable regions, hv's, or complimentarity determining regions, CDR's. These are loops of *ca.* half a dozen amino acids, and there are three such regions in each the H and the L chain. These hypervariable loops, although separated in sequence, come together in space in the folded tertiary structure the immunoglobulin assumes in solution. The resulting six proximate, hypervariable loops define the fine specificity of the immunoglobulin in question. Great structural diversity can be generated at the genetic level, and the antibody repertoire is enormous (4). Immunoglobulin hinge regions, located at the junction of the second and third domain of each H chain are susceptible to enzymatic cleavage. Reduction, followed by papain or pepsin treatment can yield Fab or Fab' fragments. Murine IgA, IgG, human immunoglobulins, *etc.*, have their inter-chain disulfide bridges in differing positions, and individual approaches should be taken towards Fab preparation. These Fab or Fab' fragments have a molecular weight of *ca.* 50,000 Daltons, and consist of the L chain and the first two domains of the H chain. They contain only *one* immunoglobulin combining area, and have affinities for the antigenic determinant identical to that of the original whole antibody (5). Their *univalence* makes them capable of binding to polymeric antigens without crosslinking and precipitating them.

MEASUREMENT OF AFFINITY BETWEEN ANTIBODY AND ANTIGEN

The noncovalent binding between an antibody's combining area (Abca) and the ligand it binds is an equilibrium reaction:

$$Abca + lig \rightleftharpoons Abca\text{--}lig$$

The equilibrium constant defines the affinity: $\quad K_a = \dfrac{C_{Abca\text{--}lig}}{C_{Abca} \cdot C_{lig}}$

The experimental measurement of the concentrations of bound antibody, free antibody, as well as the free ligand concentration can be done by several methods. Obtaining these data then yields the value for the K_a. In our laboratory we generally use either ligand-induced fluorescence change of the antibody, or micro-calorimetry. In the former, the method we employ uses the incremental change in the antibody's own tryptophanyl fluorescence as a function of ligand aliquots added (6,7; the carbohydrate ligands themselves are transparent to UV light and do not show any fluorescence). The method has been verified by comparison of the derived affinity constant with those obtained by other methods (8, 9, 10). Its accuracy is far greater than for results obtained by equilibrium dialysis, especially when the values for affinity are low.

The maximal ligand-induced change in protein fluorescence is an intrinsic property of a particular immunoglobulin, and appears to be related to the number of perturbable tryptophanyl residues in the immunoglobulin combining area. The computation (7) of the affinity constant, K_a, uses each change in protein tryptophanyl fluorescence, ΔF, caused by a given concentration of ligand and divides this by the maximally attained fluorescence change, ΔF_{max}, at infinite ligand concentration when all antibody sites are occupied by ligand. This $\Delta F/\Delta F_{max}$, which we call v, defines the fraction of antibody sites possessing bound ligand for given concentrations of free ligand. The fractional concentration of free antibody, of course, becomes $(1-v)$:

$$K_a = \frac{C_{Abca\text{--}lig}}{C_{Abca} \cdot C_{lig}}$$

i.e.
$$K_a = \frac{v}{(1\text{-}v) \cdot C_{lig}}$$

or
$$v/C_{lig} = (1\text{-}v) \, K_a$$

Concentrations are expressed as molarity, and a Scatchard plot of v/C_{lig} *versus* v yields the K_a as the intercept on the ordinate.
In measuring the affinty of *polymeric antigens* , in this case a polysaccharide, with its antibody, the question becomes: what to use for the C_{lig} ? This, because the ligand is now a polysaccharide, possessing multiple determinants (in addition the polysaccharide possesses a non-uniform molecular weight). Is the molecular weight to be used that of the determinant segment in the chain or the (number average) molecular weight of the polysaccharide? Taking into account other considerations (see below), this is precisely how antibodies that bind *internal* sequences can be distinguished from those binding a *terminal* sequence of the antigen only. In the latter case, for unbranched polysaccharides, the concentration to be used for the antigen must be the weight per liter divided by the number average molecular weight of the entire polysaccharide. Thus the concentration (that is, the availability) of the antigenic determinant is one per polymeric chain. In the former case, where the antibody can bind to internal residues, the "concentration" (or availability) is that of sequential segments of the chain, all potentially capable of binding an antibody combining area. There must be certain limitations in that former case, because the width of the front of the H/L interface of the antibody molecule is known to be *ca.* 40Å (11), and the length of most immunodeterminants is *ca.* 20Å (3). In addition, even though polysaccharides may possess segmental conformational stability, their flexibility - when viewed over the entire polymeric chain - is probably very considerable, so parts of the polymeric antigen chain may be spacially inaccessible to antibody binding due to coiling.
It is important to remember that - provided no change in the antigenic determinant's conformation occurs - the affinity an antibody has for the entire antigenic determinant for which it is specific, *is an innate property of that antibody*. That is to say, the antibody combining area has an affinity for the maximally

binding oligomeric, antigenic determinant, that is essentially the same wherever that determinant may occur in a polymeric chain, flanked by other determinants or not. Thus, once the affinity of the optimally binding ligand is known, measurement of the affinity using polymeric, whole antigen, and the Fab (or Fab') antibody fragment (to prevent possible precipitation during the measurement) must yield a value for the affinity that is the same. That identity is arrived at by using the appropriate concentration for C_{lig} in the computation of the K_a, and that in turn reveals if the antibody can read the determinant anywhere, or is limited to reading it at the antigenic terminus.

Any method for measuring the antibody-antigen affinity may be used, as all computations from the data must - of necessity - involve the use of value for the concentration of the determinant, C_{lig}.

Below are two examples.

ANTIDEXTRAN ANTIBODY SPECIFIC FOR THE DEXTRAN CHAIN TERMINUS ONLY

IgA W3129 (8, 9, 12-16) is a murine immunoglobulin specific for tetrasaccharide fragments of $\alpha(1{\to}6)$-glucopyranans (dextrans). It was shown to be precipitated by branched-, but not by linear dextrans (9). It was also shown that, in the binding of the antibody subsite having the highest affinity for a glucosyl moiety, hydrogen bonding of each the 4- and 6-OH in the glucosyl moiety was critical (14, 15). The only glucosyl moiety in an $\alpha(1{\to}6)$-dextran chain possessing a 6-OH group is - of course - the terminal residue. Thus, that finding strongly suggest that the antibody can bind to the tetrasaccharide of the dextran terminus only. A molecular fitting for the chain terminus into the antibody combining area has been proposed (15). The IgA has a maximal affinity for methyl α-isomaltotetraoside, showing a $K_a =$ 1.8 x 10^5 L/M (14). The fluorescence change of the IgA W3129 Fab' with a linear, fractionated, synthetic $\alpha(1{\to}6)$-dextran (17) having a molecular weight of ~ 36,000 Daltons was measured as a function of polysaccharide concentration. In the Scatchard plot of v/C_{lig} *versus* v , the use of

$$C_{lig} = \frac{\text{weight of dextran per Liter}}{36,000}$$

led to a K_a value of 2.0 x 10^5 L/M, *i.e.* the correct, previously found affinity found for the antibody with the tetrameric determinant (8). Hence, the antibody read one determinant per chain, and it was thus confirmed to bind the tetra-isomaltosyl fragment at the chain-terminus only.

ANTIGALACTAN ANTIBODY CAPABLE OF BINDING INTERNAL GALACTOSYL UNITS

It was previously shown (18, 19) that murine IgA J539 is specific for $\beta(1\rightarrow6)$-linked galactopyranosyl units. The immunoglobulin monomer showed (20) a precipitin line on Ouchterlony double diffusion *versus* a $\beta(1\rightarrow6)$-linked galactopyranan from *Prototheca Zopfii* having occasional, single side residues of $\beta(1\rightarrow3)$-linked galactofuranosyl residues (21, 22). This strongly suggested its capability to bind internal residues of $\beta(1\rightarrow6)$-linked galactopyranose (it was shown that IgA J539 did not bind to methyl β-galactofuranoside). Confirmation of that came in binding studies when it was shown that the immunoglobulin could bind oligomeric ligands of $\beta(1\rightarrow6)$-linked galactopyranose having their galactosyl residue(s) flanked by gentiobiose units, the latter being incapable of binding the antibody (23). The affinities observed were nearly identical to those for the $\beta(1\rightarrow6)$-linked oligomers of galactopyranose lacking the flanking residues of gentiobiose. The maximally binding ligand for IgA J539 is methyl $\beta(1\rightarrow6)$-tetragalactopyranoside (19) for which the antibody shows a K_a of 5.9 x 10^5 L/M. The galactan from *P. Zopfii* (21, 22) used in binding studies with IgA Fab' J539 had a molecular weight of ~2 x 10^5 as deduced from its K_{av} on Sephadex G-200 (20).

The IgA J539 Fab' was titrated with the galactan from *P. Zopfii*, and its polysaccharide-induced antibody fluorescence change was monitored as a function of antigen additions. In the Scatchard plot of v/C_{lig} *versus* v, the known antibody affinity constant of K_a = 5.9 x 10^5 L/M was computed, using for the ligand concentration

$$C_{lig} = \frac{\text{weight of polysaccharide per Liter}}{M_r \text{ of 30 galactosyl residues}}$$

where M_r stands for the relative molecular mass.

Thus, the antibody reads the polysaccharide antigen - having a molecular weight of ~2 x 10^5 Daltons, or *ca.* 1200 galactosyl residues per chain - as having stretches of ~30 galactosyl residues available for binding. Thus the measurement reveals the antibody to be capable of binding internal segments of the antigenic chain. It is worthwhile noting that 30 residues represent a length of some 150Å. The antibody H/L interface is about 40Å wide (11), so on the average the antigenic chain does not appear to permit closest packing with antibody combining areas. That is not surprising, as the high flexibility of the $\beta(1\rightarrow6)$-linkage quite likely allows a high degree of coiling in the antigen molecule, at any given time making parts of the chain unavailable for possible antibody-binding.

In summary: The use of Fab or Fab' fragments allows the quantitative evaluation of the affinity the antibody has for *whole* antigens, and the method has also been used for the study of other antigens, such as phosphorylcholine-bearing immunogens (24).

LITERATURE CITED

1. Kabat, E. A. *Structural Concepts in Immunology and Immunochemistry* 2nd edition, 1976, Holt Rinehart and Winston, New York.
2. Glaudemans, C. P. J.; Kováč P. *Fluorinated Carbohydrates. Chemical and Biochemical Aspects.* ACS *Symposium Series*, 1988, **374**, 78.
3. Glaudemans, C. P. J. *Chem. Rev.*,1991, **91**, 25.
4. Abbas, A. K.; Lichtman, A. H.; Pober, J. S. *Cellular and Molecular Immunology*, 1991, W. B. Saunders, Philadelphia.
5. Jolley, M. E.; Rudikoff, S.; Potter, M; and Glaudemans, C. P. J. *Biochemistry*, 1973, **12**, 3039.
6. Jolley, M. E.; Glaudemans, C. P. J. *Carbohydr. Res.*, 1974, **33**, 377.
7. Glaudemans, C. P. J.; Jolley, M. E. *Methods Carbohydr. Chem.*, 1980, **VIII**, 145.
8. Bennett, L. G.; Glaudemans, C. P. J. *Carbohydr. Res.*, 1979, **72**, 315.
9. Cisar, J; Kabat, E. A.; Dorner, M. M.; Liao, J. *J. Exp. Med.* 1975, **142**, 435.
10. Zidovetzky, R.; Blatt, Y.; Glaudemans, C. P. J.; Manjula, B. N.; Pecht, I. *Biochemistry*, 1980, **19**, 2790.
11. Padlan, E. A.; Segal, D. M.; Spande, T. F.; Davies, D. R.; Rudikoff, S; Potter, M. *Nature*, 1973, **145**, 165.
12. Weigert, M.; Raschke, W. C.; Carson, D.; Cohn, M. *J. Exp. Med.*, 1974, **139**, 137.
13. Cisar, J.; Kabat, E. A.; Liao, J.; Potter, M. *J. Exp. Med.*, 1974, **139**, 159.
14. Glaudemans, C. P. J.; Kováč, P.; Rao, A. S. *Carbohydr. Res.* 1989, **190**, 267.
15. Nashed, E. M.; Perdomo, G. R.; Padlan, E. A.; Kováč P.; Matsuda, T.; Kabat, E. A.; Glaudemans, C. P. J. *J. Biol. Chem.*1990, **265**, 20699.
16. Borden, P.; Kabat, E. A. *Proc. Natl. Acad. Sci. U.S.A.*, 1987, **84**, 2440.
17. Ruckel, E. R.; Schuerch, C. *Biopolymers*, 1967, **5**, 515.
18   Potter, M; Glaudemans, C. P. J. *Methods Enzymol.*, 1972, **28**, 388.
19. Glaudemans, C. P. J. *Mol. Immunol.*, 1987, **24**, 371.

20. Glaudemans, C. P. J.; Bhattacharjee, A. K.; Manjula, B. N. *Molec. Immunol.* 1986, **23**, 655.
21. Manners, D. J.; Pennie, I. R.; Ryley, J. F. *Carbohydr. Res.*, 1973, **25**, 186.
22. Jack, W.; Sturgeon, R. J. *Carbohydr. Res.*, 1976, **49**, 335.
23. Ziegler, Th.; Sutorius, H.; Glaudemans, C. P. J. *Carbohydr. Res.* in press.
24. Glaudemans, C. P. J.; Manjula, B. N.; Bennett, L. G.; Bishop, C. T. *Immunochemistry*, 1977, **14**, 675.

RECEIVED April 9, 1992

Chapter 7

Purification of Oligosaccharide Antigens by Weak Affinity Chromatography

D. A. Zopf[1] and W-T. Wang[2]

Biocarb Inc., 300 Professional Drive, Gaithersburg, MD 20879

Weak affinity ($K_a \approx 10^2\text{-}10^4$ M^{-1}) monoclonal antibodies immobilized at high concentrations (> 50 mg/mL) on macroporous silica matrices provide a chromatographic medium for high performance separation of oligosaccharide antigens. Unlike classical affinity chromatography, weak affinity chromatography utilizes dynamic equilibrium kinetics as a basis for separation of antigens. Simple calculations based upon established principles of chromatographic theory enable rapid determination of affinity constants and systematic adjustment of column parameters to optimize analytical and semipreparative chromatography. Demonstrated applications of weak affinity chromatography include analysis of picomol quantities of oligosaccharides in complex biological fluids such as urine and serum and recovery of oligosaccharide antigens released from partially fractionated oligosaccharides of pooled human milk.

Antibodies with weak affinity ($K_a < 10^4$ M^{-1}) are abundant among natural circulating immunoglobulins and are frequently observed as products of B-cell hybridomas, but they have received little attention either as biological mediators or as potential immune reagents. Indeed, it has long been assumed that triggering of affinity maturation not only reflects the most significant aspect of immune adaptation, but also provides the best reagents for immunoassays. Recently a "dynamic" form of affinity chromatography was introduced which exploits the advantages inherent in weak affinity interactions as a basis for continuous chromatographic separations under isocratic conditions (1, 2). Termed "weak affinity chromatography" (WAC), this technique has been pioneered using antibody-oligosaccharide interactions and is especially well-suited for studying effects of structural modifications of oligosaccharide antigens on immune recognition.

WAC: Theoretical Basis

Basic chromatographic theory is well-established in a long series of publications (see, e.g., refs. 3, 4). For a ligate, S, dissolved in the bulk phase in a WAC system and interacting with an immobilized ligand, P, it may be reasonably assumed that in most cases binding equilibrium will occur via a second order, reversible reaction:

[1]Current address: 3252 Patterson Street, N.W., Washington, DC 20015
[2]Current address: MicroCarb Inc., 300 Professional Drive, Gaithersburg, MD 20879

$$S + P \underset{\longleftarrow}{\overset{K_a}{\longrightarrow}} SP \qquad (1)$$

where SP is the ligand/ligate complex and K_a the association constant (M^{-1}). The Langmuir adsorption isotherm for the system can be expressed as:

$$\sigma = \frac{P_{max}K_aS'}{1 + K_aS'} \qquad (2)$$

where σ is the concentration of adsorbed ligate at equilibrium, S' is the concentration of free ligate at equilibrium, and P_{max} is the maximum accessible ligand sites. This expression approaches linearity for values of $K_aS' \ll 1$. Introduction of a constant, C (usually in the numerical range 0.5 - 1) that describes features of a chromatography column such as porosity, void fraction, and density of the stationary phase, enables simple calculation of ligate retention expressed as the capacity factor, k', according to the equation:

$$k' = C P_{max}K_a \qquad (3)$$

When chromatography is performed under conditions where $K_aS' \ll 1$, *i.e.*, within the linear region of the adsorption isotherm, kinetic contributions to peak broadening are negligible, so that chromatographic performance of affinity chromatography under optimal conditions is predicted to approach that associated with high performance liquid chromatography using familiar reverse phase or ion-exchange adsorbents (1). From equation 3 it could be predicted, for example, that k' =1 might be achieved with high performance if an IgG monoclonal antibody interacting with an oligosaccharide with K_a = 10^3 M^{-1} were immobilized at a concentration of 10^{-3} M ($\approx$ 75 mg/mL) on a macroporous silica bed (C $\approx$ 1) provided that the concentration of the oligosaccharide sample is $\leq 10^{-5}$ M.

Applications of WAC

Influence of Temperature on K_a. Temperature dependent changes in affinity between monoclonal antibodies and oligosaccharides have been observed repeatedly (5, 6, 7). For example, Lundblad *et. al.* (5) noted that antibody 61.1 (IgG3) decreased in affinity for a glucose containing tetrasaccharide, $(Glc)_4$, by a factor of 2 for each 8° rise in temperature from 4° to 50° C. A second monoclonal antibody, 39.5 (IgG2b) exhibits a similar rate of change of affinity over the same temperature range (Figure 1). Monoclonal antibody A003 (IgM) (6) that binds blood group A-active oligosaccharides and antibody CO 514 (IgG3) that binds Le[a]-active oligosaccharides (7) exhibit similar slopes for K_a *vs* temperature, although they vary with respect to the value of K_a at any given temperature. The mechanism(s) responsible for these changes in affinity have not been fully explored. While not all monoclonal antibodies that bind these or other oligosaccharides display this same degree of temperature dependence (W-T Wang and D. Zopf, unpublished observations), the observation that affinity can vary substantially, and reversibly, over the physiological temperature range provides important opportunities for manipulating affinity in many WAC systems.

Modulation of Chromatographic Performance. WAC systems where K_a for the ligand/ligate complex has a relatively strong temperature dependence provide an

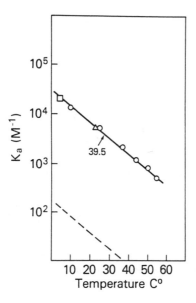

Figure 1.   Affinity constant (K_a) as a function of temperature for binding of the oligosaccharide (Glc)$_4$ by monoclonal antibody 39.5 . Experimental points were determined by the following methods: o equilibrium dialysis; Δ nitrocellulose filtration; $\square$ frontal analysis. The dashed line represents a hypothetical cold agglutinin.

excellent demonstration of the variation in chromatographic performance with K_a. For example, monoclonal antibody 39.5 was immobilized on tresyl-activated ten micron macroporous silica beads (SelectiSpher 10, Perstorp Biolytica) at 86 mg/mL and 55% of calculated active sites were determined by frontal analysis to retain antigen binding activity (1). Affinity chromatography of the tritiated oligosaccharide alditol, $[^3H](Glc)_4$-ol, at various temperatures over the range 10° - 50° C (Figure 2) illustrated a dramatic increase in chromatographic performance at weaker affinities as predicted from theory (see above).

Analytical WAC. The same affinity column was used to analyze for $(Glc)_4$ excreted in human urine (Figure 3) providing a rapid, highly sensitive immunoassay system where specificity arises from a multitude of cumulative weak interactions resulting in sharp retarded peaks under isocratic conditions, and sensitivity is provided by selective electrochemical detection of the analyte (8). The same analytical system has simplified routine clinical investigations of the excretion of $(Glc)_4$, a limit dextrin from amylolysis of glycogen (9), in blood and urine of patients with acute pancreatitis (8).

Affinity separation of three blood group A-active oligosaccharides provides a further illustration of the use of temperature to adjust K_a in order improve chromatographic performance. Three oligosaccharides known to bind antibody A003 with differing affinities (10) were chromatographed on a column containing the IgM antibody non-covalently bound to Selectispher 10-Con A (6). Non-covalent binding of immunoglobulins to affinity matrices prevents loss of binding activity due to unwanted chemical attack by the coupling reagent at or near antibody binding sites (11) and enables easy recovery and reuse of both the antibody and column. Figure 4b demonstrates the improvements in chromatography that can be easily obtained by stepping up column temperature to increase the elution rates and improve the peak shapes of higher affinity oligosaccharide species (Figure 4a).

Semipreparative WAC

The use of WAC as a semipreparative method has many advantages over classical chemical methods for purification of oligosaccharides from complex mixtures. Antibodies that recognize a known "immunodominant" carbohydrate epitope can selectively retard oligosaccharides that carry this epitope. Moreover, the affinity for oligosaccharides may vary when the epitope is built up on core structures that differ in composition and complexity (7,10). As a result, WAC may be used to isolate oligosaccharides that share a common epitope from mixtures containing many compounds of similar composition and molecular weight, and in some cases may completely or partially resolve components that share a common epitope attached to different cores.

Purification of Known Antigens. Isolation of the Le^a-active oligosaccharide, lacto-N-fucopentaose II (LNF II) (see Table I for oligosaccharide structures), by affinity chromatography from a complex mixture of neutral oligosaccharides from pooled human milk is illustrated in Figure 5. In this experiment, LNF II is isolated as a single retarded peak under isocratic conditions from more than twenty oligosaccharides with related chemical structures, including an isomeric pentasaccharide, lacto-N-fucopentaose III (7). Scale-up of this isolation procedure to rapidly isolate milligram quantities of pure pentasaccharide would provide a favorable alternative to the multiple chromatographic steps required for isolation by paper chromatography (12) or HPLC (13). On the other hand, industrial scale preparative WAC for production at the kilogram level, although feasible in principle, will await further development of technologies for producing weak affinity ligands, such recombinant VH fragments, at low cost.

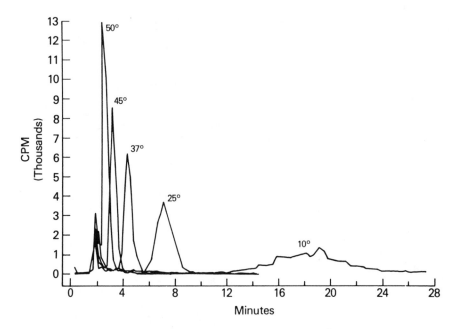

Figure 2.   Weak affinity chromatography of [3H](Glc)$_4$-ol as a function of temperature. Identical samples (1.5 ng; sp. act. = 8.4 Ci/mmol) injected in 10 μL distilled water were run isocratically in 0.2M NaCl plus 0.02 M sodium phosphate buffer, pH 7.5 and fractions counted by liquid scintillation. (Reproduced with permission from ref. 1. Copyright 1988 Academic Press.)

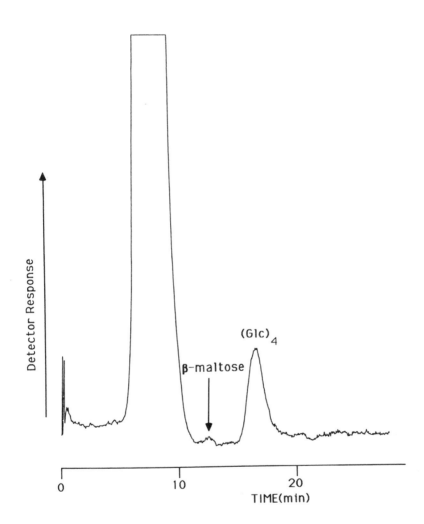

Figure 3.   Weak affinity chromatographic analysis of (Glc)₄ in normal human urine. Ultrafiltered and deionized urine (25 μL) was injected onto the same column as was used in Figure 2 and the system was eluted with 0.1M Na_2SO_4 plus 0.02 M phosphate, pH 7.5 at 30° C. The column effluent was mixed with 50mM NaOH prior to passage through a pulsed amperometric detector [see (8) for further details]. (Reproduced with permission from ref. 8. Copyright 1989 Academic Press.)

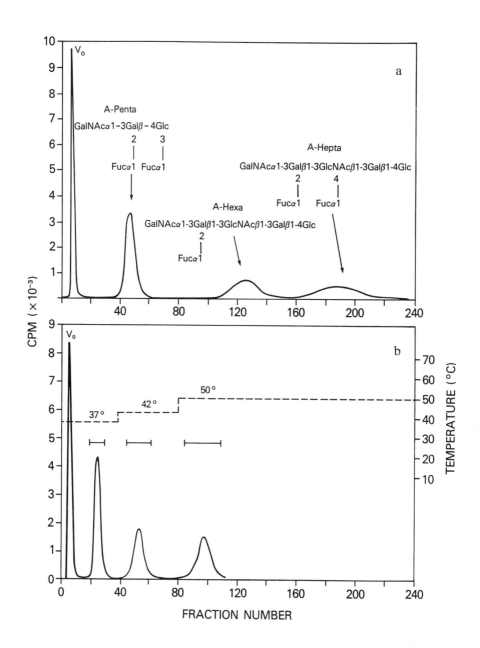

Figure 4.   Affinity chromatography of three blood group A-active
oligosaccharides on a 0.5 X 10 cm column containing 16 mg antibody A003 (IgM)
non-covalently bound to SelectiSpher 10-Con A (Perstorp Biolytica). (a) isocratic
elution at 25° C; (b) elution at 37° C followed by step-wise temperature increases
as shown [see (6) for further details]. (Reproduced with permission from ref. 6.
Copyright 1987 Academic Press.)

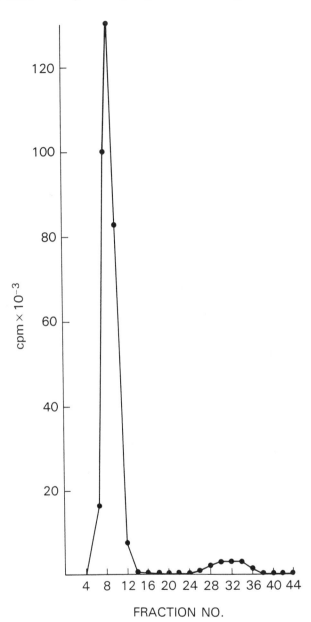

Figure 5.   Affinity isolation of LNF II from a mixture of human milk oligosaccharides. A mixture of neutral di- through hexasaccharides from pooled human milk (1.2 X 10^{-2} nmol) was applied in 50μL to a 0.3 X 10 cm glass column containing 9.5 mg monoclonal antibody CO 514 (IgG3) non-covalently immobilized on *Staph*. protein A-Sepharose (Sigma Chemical Co.). The system was run isocratically in 0.14 M NaCl plus 0.05 M Tris-HCl, pH 8.0 at 0.1 ml/min, 25° C [see (7) for further details]. (Reproduced with permission from ref. 7. Copyright 1988 Academic Press.)

Table I. Structures of Oligosaccharides

Name	Structure
(Glc)$_4$	Glcα1-6Glcα1-4Glcα1-4Glc
lacto-N-fucopentaose II	Galβ1-3GlcNAcβ1-3Galβ1-4Glc 4 | Fucα1
lacto-N-fucopentaose III	Galβ1-4GlcNAcβ1-3Galβ1-4Glc 3 | Fucα1

Isolation of Unknown Antigens for Structural Characterization. Separation of three blood group A-active oligosaccharides that share the non-reducing trisaccharide epitope, GalNAcα1-3(Fucα1-2)Galβ1- , but differ in the composition and complexity of their core structures, was illustrated in Figure 4. This experiment suggests a simple, general approach for identifying the native carbohydrate receptor molecule recognized by an antibody, or other lectin-like molecule: oligosaccharides released intact from glycoconjugates by endoglycanases or chemical procedures can be chromatographed on columns substituted at high density with monoclonal antibodies or other recombinant lectin-like molecules which are relatively easy to prepare in quantities appropriate for use as ligands in analytical or semipreparative WAC. This approach provides a convenient alternative to searching for an oligosaccharide inhibitor, a commonly used experimental device for defining the specificity of molecules that bind oligosaccharide targets on cells and tissues.

For example, investigations of the fine specificity of a monoclonal antibody that recognizes a cancer-associated antigen on mucins circulating in the serum of many patients with breast cancer has been carried out using WAC (14). Linsley, et al (15) noted that the antibody, Onc-M26, reacts with a ganglioside containing six sugars bearing the sialyl-Le[x] (SLe[x]) antigen, NeuAcα2-3Galβ1-4(Fucα1-3)GlcNAcβ1-3Galβ1-4Glcβ1-Cer; however, immunostaining of the total gangliosides from breast cancer tissue also revealed intense staining of a higher molecular weight, trace component, suggesting that the avidity of Onc-M26 was significantly greater for a more complex carbohydrate moiety. WAC of oligosaccharides from human milk, a rich source of sialylated and fucosylated oligosaccharides (12), resulted in isolation of two SLe[x]-active nonasaccharides (Figure 6) with K_a's for Onc-M26 approximately twenty times greater than that of the SLe[x]-active hexasaccharide present in the major SLe[x]-active breast cancer ganglioside. Onc-M26 is an IgM antibody, potentially capable of decavalent binding. Its *avidity* for polyvalent antigen, such as aggregated gangliosides or a multiply glycosylated mucin molecule, could increase by several logs as a result of this 20-fold increase in *affinity* at single sites involved in cooperative binding.

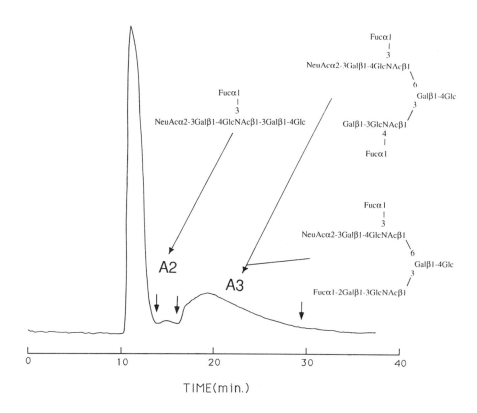

Figure 6. Affinity isolation of SLe^x-active oligosaccharides. A sample containing 1 μg partially fractionated monosialylated oligosaccharides from pooled human milk was run isocratically at 0.2 mL/min with 0.2 M Na_2SO_4 plus 0.2 M phosphate buffer pH 7.5 at 0 °C. NaOH (0.1 M) was added via a post column mixing tee, and oligosaccharides were detected with a pulsed amperometric detector. Fractions A2 and A3 were pooled separately as indicated by arrows. Fraction A3 was further fractionated by high-performance ion-exchange chromatography giving two major, well-separated peaks that were isolated and shown by NMR to have the chemical structures indicated. (Reproduced with permission from reference 8. Copyright 1989 Academic Press.)

Weak Affinity in Biological Systems

Cold Agglutination - A Hypothesis. Specific recognition of oligosaccharides astargets for hemagglutination forms the basis for the ABO, Lewis, P, Ii, and other human blood groups (16). Antibodies that bind bacterial polysaccharides provide the basis for serotyping a vast array of microbes. Anti-carbohydrate immune recognition in normal healthy animals and humans is commonly mediated by antibodies molecules whose individual combining sites bind antigen with relatively weak affinity ($Ka \leq 10^4$ M^{-1}); however, interactions at multiple sites simultaneously can provide adhesive forces of sufficient strength to mediate stable cell to cell contact, activation of complement, and other immune phenomena.   The fairly steep dependence of affinity on temperature across the range $4°$ - $37°$ observed for many murine monoclonal antibodies invites speculation as to the cause for the common phenomenon of IgM cold autoagglutinins in human and animal sera (17). It seems reasonable to assume that carbohydrate antigens in the environment stimulate some B-cells to produce antibodies cross-reactive with red blood cell antigens, but with affinities too weak to mediate binding at $37°$ C. Such antibodies do not mediate immune recognition of self antigens at body temperature, and therefore are "permitted". A subset of these antibodies might exhibit enhanced affinity for oligosaccharide antigens at colder temperatures in a manner similar to that reported for murine monoclonal antibodies (see above). If on the average, three clustered antigen molecules were available for simultaneous binding on each red cell surface (18), a shift in single-site affinity from 10^1 M^{-1} at $37°$ to 2 X 10^2 M^{-1} at $4°$ might result in a change in avidity from approximately 10^3 M^{-1} at the warmer temperature to approximately 10^7 M^{-1} in the cold. Malignant transformation of individual B-cell clones that produce such antibodies could give rise to monoclonal cold agglutinin disease in humans with Waldenstrom's macroglobulinema.

In a wider context, weak affinity binding of oligosaccharides by immunoglobulins can be thought of as a model for recognition of cell surface glycoconjugates by microbial and mammalian lectins. It is now widely accepted that microbial pathogens specifically attach to cells (19) and cells specifically aggregate with each other (20) and interact with some components of the extracellular connective tissue matrix via a multitude of weak affinity interactions that must be readily reversible in order to permit a dynamic environment suited to constant adaptation. Weak affinity chromatography provides a powerful new approach to isolation, detection, and kinetic analysis of the components of many of these recognition systems.

Acknowledgments

We wish to thank our many colleagues, especially Professor Arne Lundblad and Dr. Sten Ohlson, whose inspiration, hard work, and many helpful discussions have contributed much to the data and ideas presented herein.

Literature Cited

1.   Ohlson, S.; Lundblad, A.; Zopf, D. *Anal. Biochem.* **1988**, *169*, 204 - 208.
2.   Zopf, D.; Ohlson, S. *Nature* **1990**, *346*, 87 - 88.
3.   Kucera, E. *J. Chromatogr.* **1965**, *19*, 237 - 248.
4.   Goldstein, S. *Proc. R. Soc. London*, **1953**, *219*, 151 - 163.
5.   Lundblad, A; Schroer, K.; Zopf, D. *J. Immunol. Methods* **1984**, *68*, 227 - 234.
6.   Dakour, J.; Lundblad, A.; Zopf, D. *Anal. Biochem.* **1987**, *161*, 140 - 143.

7. Dakour, J.; Lundblad, A.; Zopf, D. *Arch. Biochem. Biophys.* **1988**, *264*, 203 - 213.
8. Wang, W-T.; Kumlien, J.; Ohlson, S.; Lundblad, A.; Zopf, D. *Anal. Biochem.* **1989**, *182*, 48 - 53.
9. Ugorski, M.; Seder, A.; Lundblad, A.; Zopf, D. *J. Exp. Pathol.* **1983**, *1*, 27 - 38.
10. Chen, H-T.; Kabat, E.A. *J.Biol. Chem.* **1985**, *260*, 13208 - 13217.
11. Gersten, D.M.; Marchalonis, J.J. *J. Immunol. Methods* **1978**, *24*, 305 - 309.
12. Kobata, A. In *Methods in Enzymology*; Ginsburg,V., Ed.; Academic Press, New York, NY, 1972, Vol. 28; pp 262 - 271.
13. Kallin, E.; Lönn, H.; Norberg, T. *Glycoconjugate J.* **1986**, *3*, 311 - 319.
14. Wang, W-T.; Lundgren, T.; Lindh, F.; Nilsson B.; Grönberg, G.; Brown, J.P.; Mentzer-Dibert, H.; Zopf, D. *Arch Biochem. Biophys.* **1992**, *292*, 433 - 441.
15. Linsley, P.S.; Brown, J.P.; Magnani, J.L.; Horn, D. *Proceedings* IXth International Symposium on glycoconjugates, Lille, France, 1987; Abstract F69.
16: Watkins, W. M. In *Advances in Human Genetics*; Harris, H.; Hirschhorn, K., Eds.; Plenum Press, New York, NY, 1980, Vol 10; pp 1 - 136.
17 Bird, G.W.G. *Blut* **1966**, *12*, 281 - 285.
18. Hoyer, L.W.; Trabold, N.C. *J. Clin Invest.* **1971**, *50*, 1840 - 1846.
19. Leffler, H.; Svanborg-Eden, C. In *Microbial Lectins and Agglutinins: Properties and Biological Activity*; Mirelman, D., Ed., John Wiley & Sons, New York, NY, 1986, pp 83 - 111.
20. Phillips, M.L.; Nudelman, E.; Gaeta, F.C.A.; Perez, M.; Singhal, A.K.; Hakomori, S-I.; Paulson, J.C. *Science* **1990**, *250*, 1130 - 1135.

RECEIVED March 25, 1992

Chapter 8

Synthesis and Conformational Analysis of the Forssman Pentasaccharide and Di-, Tri-, and Tetrasaccharide Fragments

G. Magnusson, U. Nilsson, A. K. Ray, and K. G. Taylor[1]

Organic Chemistry 2, Chemical Center, The Lund Institute of Technology, University of Lund, P.O. Box 124, 221 00 Lund, Sweden

The biological importance of the Forssman antigen and some of its oligosaccharidic fragments is reviewed as is the synthesis of these fragments. Synthesis of the di—pentasaccharides β-D-GalNAc-(1-3)-α-D-Gal-1-OMe, α-D-GalNAc-(1-3)-β-D-GalNAc-(1-3)-α-D-Gal-1-OMe, β-D-GalNAc-(1-3)-α-D-Gal-(1-4)-β-D-Gal-1-OTMSEt, α-D-GalNAc-(1-3)-β-D-GalNAc-(1-3)-α-D-Gal-(1-4)-β-D-Gal-1-OTMSEt, β-D-GalNAc-(1-3)-α-D-Gal-(1-4)-β-D-Gal-(1-4)-β-D-Glc-1-OTMSEt, and α-D-GalNAc-(1-3)-β-D-GalNAc-(1-3)-α-D-Gal-(1-4)-β-D-Gal-(1-4)-β-D-Glc-1-OTMSEt, are briefly described. The use of X-ray crystallography, NMR spectroscopy, and computer methods for conformational analysis of Forssman-related saccharides is discussed.

Antigens of the globoseries of glycolipids (Figure 1) in conjunction with the various proteins that bind to the oligosaccharide portions of these antigens constitutes a useful model system for the study of molecular recognition between carbohydrate and protein. These antigens are recognized *in vivo* by antibodies of the P blood-group system and by various bacterial proteins such as the pilus-associated PapG adhesin protein of *Escherichia coli* (*1*) as well as verotoxin (*2*) (from *E. coli*) and Shiga toxin (*3*) (from *Shigella dysenteriae*). Furthermore, glycolipids of the globoseries have been suggested to play important roles as tumor-associated antigens on Burkitt lymphoma cells (*4*) and are also enriched in body fluids of patients suffering from Fabry's disease (*5*). The present paper is concerned with the synthesis and conformation of saccharides related to the Forssman pentasaccharide (Figure 1) with special reference to the work done on adhesion of pathogenic *E. coli* bacteria to human epithelial cells of the urinary tract.

[1]Current address: College of Arts and Sciences, University of Louisville, Louisville, KY 40292

0097–6156/93/0519–0092$06.00/0
© 1993 American Chemical Society

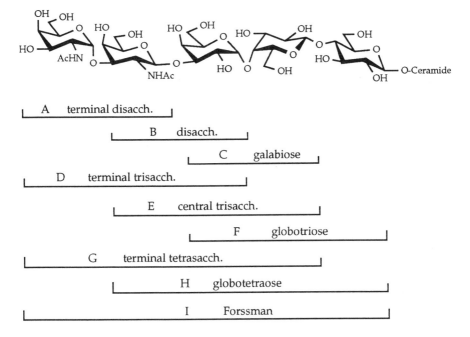

Figure 1. Structure of the Forssman antigen and assignment of the fragments discussed in the paper.

The Pap Proteins of *E. coli.*

Uropathogenic *E. coli* bacteria carry different filamentous protein aggregates anchored
to the bacterial surface. These aggregates are termed pili (or fimbriae) and the Pap-pili
(Pap = pyelonephritis associated pili) consist of approximately 1000 helically arranged
subunit proteins (Figure 2). Recombinant DNA experiments (*1*) have shown that the
Pap-pili are polymers of the major subunit PapA and the minor subunits PapE, PapF,
and PapG. The latter contribute only one or a few copies to each pilus. Deletion of the
PapA gene inhibited pilus formation but not the potential for adhesion. The minor Pap
subunits are essential for adhesion and were shown to be located at the pilus tip. PapG
was shown to be the adhesin protein that binds the bacteria to galabiose-containing
glycolipids on cell surfaces; each pilus seems to contain only one PapG molecule. A
so called chaperonin protein (PapD) was shown to be important for folding and trans-
periplasmic transport of the other Pap proteins to the site of pilus assembly (*6*). The
crystal structure of PapD was recently reported (*7*).
Over-expression of PapG in genetically altered *E. coli*, followed by affinity
purification on a galabiose-containing Sepharose gel (*6*) has so far produced several
milligrams of a pure PapG/PapD complex to be used in crystallography and adhesion
experiments. It was also shown that 6 M urea caused denaturation of the complex into
nonadhesive PapG and PapD. Addition of native PapD caused refolding of PapG into
the adhesive state. Furthermore, a Glu→His point mutation in PapD (the chaperonin)
resulted in defect pili that were bent and even broken (Hultgren, S. J. Washington
University School of Medicine, St. Louis, personal communication, 1990). It should
be stressed that this dramatic event took place even though PapD is not incorporated
into the pilus but merely helps in the assembly process!

The Galabiose-Specific Receptor Site of the PapG Adhesin.

Following the original observations (*8*) that galabiose-containing saccharides inhibit
the agglutination between uropathogenic *E. coli* bacteria and human cells (red blood
cells and urine sediment cells) it was shown that the presence of galabiose in natural
glycolipids is sufficient for bacterial binding in an over-lay experiment on thin layer
plates (*9*).

Detailed information about the molecular basis for the adhesion was obtained from
an inhibition study with deoxygalabiose analogs, using an *E. coli* strain
(HB101/pPAP5) that carries the Pap pili encoded by the plasmid pPAP5 (*10*). It was
thus indicated that [i] galabiosides interact with the PapG adhesin by hydrogen bonding
involving five out of seven hydroxyl groups (HO-6, 2', 3', 4', 6'), [ii] that HO-3 was
intramolecularly hydrogen bonded to the ring oxygen O-5' (as in the crystal of
galabiose (*11*)), [iii] that HO-2 is in contact with the surrounding water, [iv] that
changing HO-3' into MeO-3' doubled the inhibitor potency, [v] that the anomeric
oxygen O-1 is not involved in hydrogen bonding to the protein, and [vi] that
hydrophobic aglycons increase the inhibitor potency. A summary and model of the
receptor interactions are shown in Figure 3.

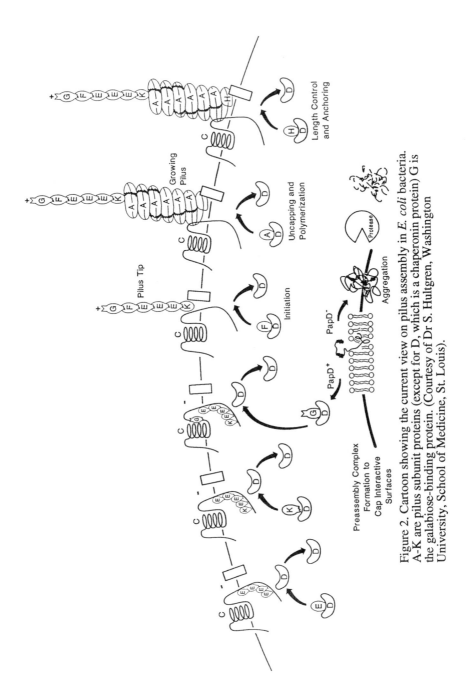

Figure 2. Cartoon showing the current view on pilus assembly in *E. coli* bacteria. A-K are pilus subunit proteins (except for D, which is a chaperonin protein) G is the galabiose-binding protein. (Courtesy of Dr S. Hultgren, Washington University, School of Medicine, St. Louis).

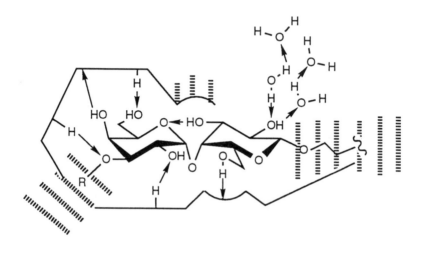

Figure 3. Schematic model of the galabiose-specific binding site of the PapG adhesin of *E. coli*. Arrows indicate hydrogen bonds between saccharide and protein and hatched areas indicate non-polar interactions. (Adapted from ref. 10).

Synthesis of the Forssman Pentasaccharide and Related Oligosaccharides.

Introduction. Since it has been indicated that different uropathogenic *E. coli* strains recognize slightly different epitopes of the glycolipids belonging to the globoseries (*12*), we decided to synthesize the Forssman pentasaccharide as well as the corresponding di-, tri-, and tetra-saccharide fragments for use in various bioassays. Our experience of anomeric protection-deprotection-activation strategies (*13*) and pre-spacer glycoside chemistry (*14*) determined the synthesis paths, which would lead to a rather comprehensive set of the globoseries glycoconjugates including soluble glycosides, neoglycolipids, neo-glycoproteins, glycoparticles, and glycosurfaces.

Numerous research groups have been involved in the synthesis of oligosaccharides of the globoseries of glycolipids and all the saccharides corresponding to the naturally occurring Forssman fragments have been synthesized. Furthermore, the complete set of monodeoxy galabiosides has been reported as well as several deoxyfluoro derivatives. However, in many cases the final products were methyl glycosides of limited use in biological research. We therefore decided to synthesize the Forssman fragments (A-I, Figure 1) as TMSEt glycosides (*13*), which permits high-yielding transformations of the complete oligosaccharides into glycoconjugates (*14*). Syntheses of the Forssman fragments of the present work as well as those reported to date are compiled in Table I.

Table I. References to Synthesis of Forssman Fragments

Saccharide fragments[a]	Glycosides and hemiacetals[b]	Glyco-conjugates[c]	Deoxy analogs[d]
A	15[e]		
B	f		
C	13, 14, 16-23	14, 24-26	27-31
D	15, f	f	
E	f	f	
F	21, 33-35	34, 36-39	
G	f	f	
H	40,41, f	41, f	
I	42, 43, f	44, f	

[a]As defined in Figure 1. [b]Soluble (e.g. methyl, allyl, benzyl) glycosides and hemiacetal sugars. [c]Glycolipids, Glycoproteins, or glycoparticles. [d]Monodeoxy- and monodeoxyfluoro analogs. [e]Protected intermediate. [f]Present work.

The use of the TMSEt-group for protection of the anomeric center of the reducing-end saccharide during synthesis of the various oligosaccharides has several advantages (*13*). First, the TMSEt group is fully compatible with most of the reaction conditions that are routinely used in glycoside synthesis, protecting group chemistry, and other saccharide modifications as exemplified in Table II.

Table II. Reaction Conditions that do not Affect the TMSEt Anomeric Blocking Group

Reaction	Reagent
Glycoside synthesis	AgOSO$_2$CF$_3$/collidine;HgO/HgBr$_2$; Hg(CN)$_2$/HgBr$_2$; Et$_4$NBr; TsOH; AgOSO$_2$CF$_3$/MeSBr; Me$_2$SSMeO$_3$SCF$_3$
Acylation	Ac$_2$O/pyridine;PhCOCl/pyridine; (CF$_3$CO)$_2$O/CF$_3$COONa; PhSO$_2$Cl/pyridine
Deacylation	MeONa/MeOH
Benzylation	PhCH$_2$Cl/KOH; PhCH$_2$Br/NaH; PhCH$_2$Br/Bu$_4$NHSO$_4$/NaOH
Hydrogenation	H$_2$/Pd/C
Acetal formation	PhCHO/HCOOH; PhCH(OMe)$_2$/TsOH; Me$_2$C(OMe)$_2$/TsOH
Acetal hydrolysis	HOAc(aq); CF$_3$COOH(aq); I$_2$/MeOH
Silylation	t-BuMe$_2$SiCl/imidazole
Desilylation	Bu$_4$NF
Regioselective allylation	Bu$_2$SnO/Bu$_4$NBr/CH$_2$=CH-CH$_2$Br
Deallylation	PdCl$_2$/MeOH
Hydride reduction	NaCNBH$_3$/HCl/Et$_2$O; LiAlH$_4$
Oxidative benzylidene cleavage	Br$_2$/NBS/CCl$_4$
Azide reduction	NaBH$_4$/NiCl$_2$/H$_3$BO$_3$
Deoxygenation	Bu$_3$SnH
Oxidation	Me$_2$SO/(COCl)$_2$
Wittig olefination	Ph$_3$P=CHCH$_3$; Ph$_3$P=CH$_2$
Nucleophilic substitution	CsOAc/DMF

Secondly, TMSEt glycosides can be transformed into the corresponding hemiacetal ("free") sugars (*13a*), 1-O-acyl sugars with conservation of the anomeric configuration (*13a*), and 1-chloro sugars (*13b*). These transformations are routinely performed in >90% yield of isolated products, even with rather sensitive compounds such as the sialyl Le[x] hexasaccharide (*45*). We feel secure enough to recommend the use of the flexible TMSEt glycosides in synthesis whenever the final oligosaccharide must be transformed into a glycolipid or other type of glycoconjugate.

Synthesis of Forssman Fragments; Present Work. The synthetic strategy was based on β-glycosylation of suitably protected TMSEt glycosides of galactose, galabiose, and globotriose with a protected galactosamine donor, followed by deprotection, reprotection and α-glycosylation with a 2-azidodeoxygalactose donor. The strategy was successful, although higher yields were obtained by block synthesis of the TMSEt glycosides in the case of globotetraose. Compound numbers refer to Chart I and II.

Synthesis of Glycosyl Acceptors.   2-(Trimethylsilyl)ethyl β-D-galactopyranoside (**1**) (*13a*) was regioselectively allylated (*46*) and then benzylated to give the 3-*O*-allyl glycoside **2** (65%). Deallylation of **2** with palladium chloride in methanol (*47*) gave the acceptor **3** (95%). Hydrolysis (*13a*) of **2**, followed by treatment with Vilsmeier reagent (*48*) gave a quantitative yield of the chloride donor **5**. Benzylidenation of **1**, followed by benzylation and reductive ring opening of the benzylidene group (*49*) gave the TMSEt galactoside acceptor **4** (*13a*) (71%).

Silver triflate-promoted glycosylation of **4** and the corresponding lactoside **6** (*13a*) with crude **5**, followed by deallylation of the intermediates **7** and **9**, gave the acceptors **8** (69%) and **10** (55%), respectively.

The GalNAcβ Linkage. Galactosides **3** and **14** (*50*), galabioside **8**, and globotrioside **10** were glycosylated with the donor **11** (*51*) using silver triflate as promoter to give **12** (70% + 9% α-glycoside), **15** (83%), **18** (75% + 5% α-glycoside), and **23** (26% + 9% α-glycoside), respectively. Thus, as reported in the literature, the larger the acceptor, the lower the yield. With the trisaccharide acceptor **10**, the yield was disappointingly low, partly due to the formation of glycal by HCl-elimination from **11**, and consequently, an alternative route was investigated. Silver triflate-mediated glycosylation of the lactoside **6** with the disaccharide donor **13** (obtained from **12** as described for **5**) gave an α/β mixture (60%), which was purified to give **23** (44%).

Deblocking of **15**, **18**, and **23** gave disaccharide glycoside **17** (77%), trisaccharide glycoside **21** (86%; after purification via the acetylated derivative **20**), and tetrasaccharide glycoside **26** (91%). Anomeric deblocking of **21** and **26**, using trifluoroacetic acid/dichloromethane (*13a*) gave the corresponding hemiacetal sugars **22** (83%) and **27** (98%).

The GalNAcα Linkage. The di-, tri-, and tetrasaccharide acceptors **16**, **19**, and **24** were prepared as follows. Benzylidenation of **15**, followed by hydrazinolysis, acetylation, and de-*O*-acetylation gave **16** (72%). Hydrazinolysis, acetylation, de-*O*-acetylation, and benzylidenation of **18** and **23** gave **19** (60%) and **24** (74%), respectively.

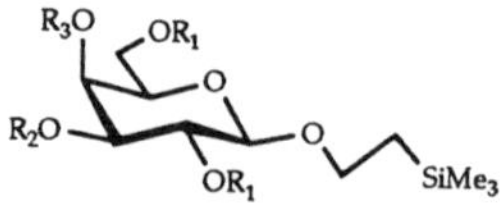

1: $R_1 = R_2 = R_3 = H$
2: $R_1 = R_3 = Bzl; R_2 = allyl$
3: $R_1 = R_3 = Bzl; R_2 = H$
4: $R_1 = R_2 = Bzl; R_3 = H$

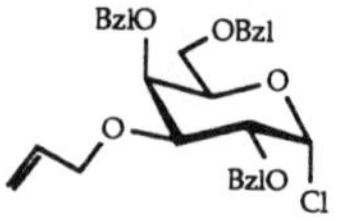

5

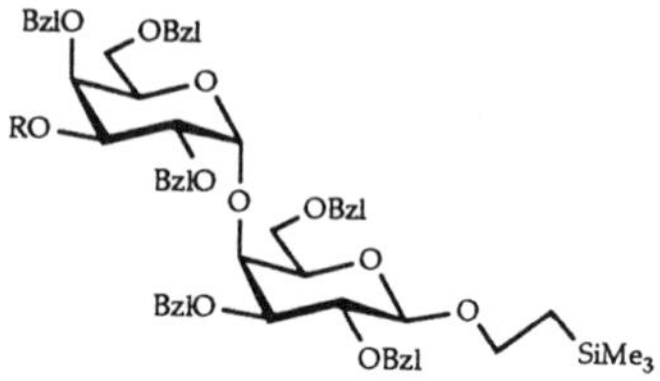

6

7: R = allyl
8: R = H

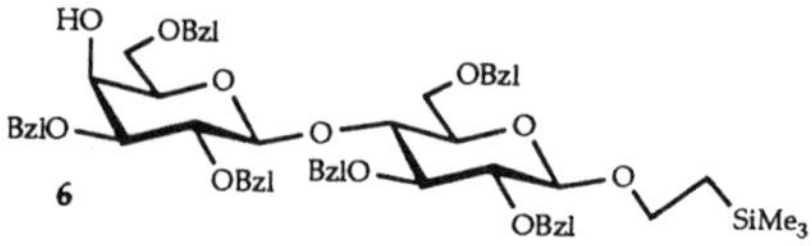

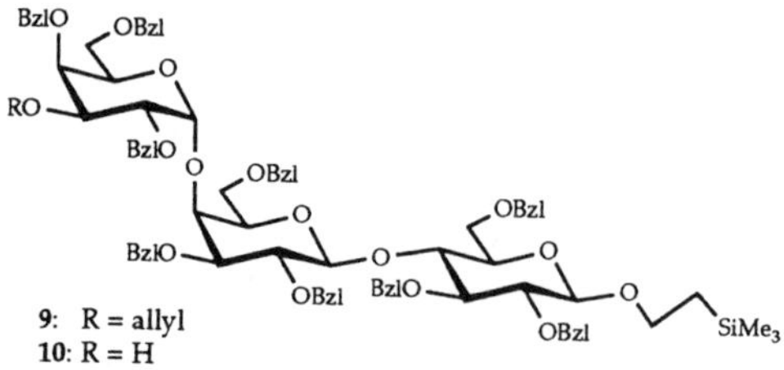

9: R = allyl
10: R = H

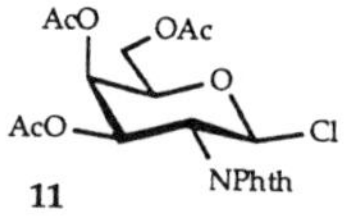

11

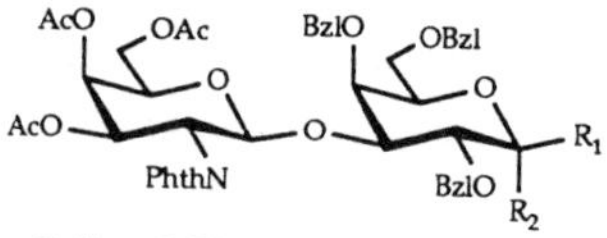

12: $R_1 = O\text{-}TMSEt; R_2 = H$
13: $R_1 = H; R_2 = Cl$

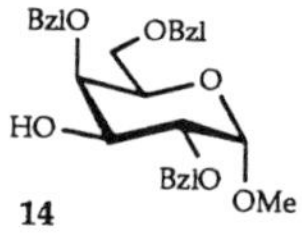

14

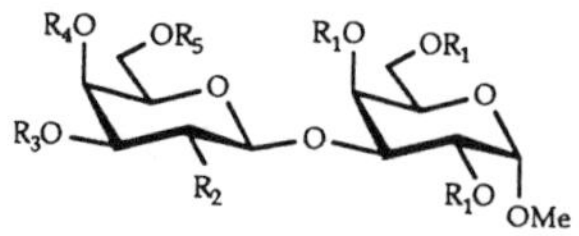

15: $R_1 = Bzl; R_2 = NPhth; R_3 = R_4 = R_5 = H$
16: $R_1 = Bzl; R_2 = NHAc; R_3 = H; R_4,R_5 = PhCH$
17: $R_1 = R_3 = R_4 = R_5 = H; R_2 = NHAc$

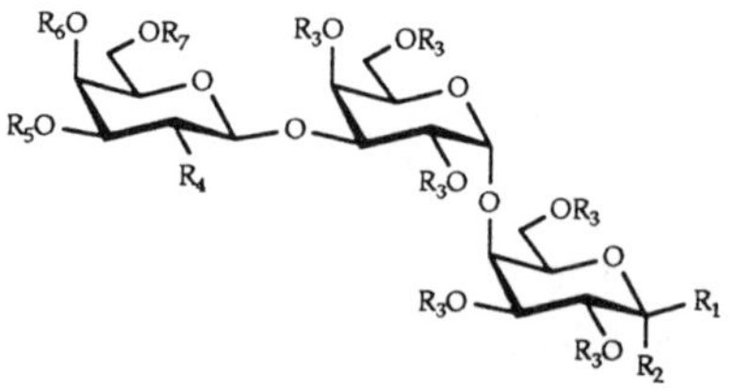

18: $R_1 = O\text{-}TMSEt; R_2 = H; R_3 = Bzl; R_4 = NPhth; R_5 = R_6 = R_7 = Ac$
19: $R_1 = O\text{-}TMSEt; R_2 = R_5 = H; R_3 = Bzl; R_4 = NHAc; R_6, R_7 = PhCH$
20: $R_1 = O\text{-}TMSEt; R_2 = H; R_3 = R_5 = R_6 = R_7 = Ac; R_4 = NHAc$
21: $R_1 = O\text{-}TMSEt; R_2 = R_3 = R_5 = R_6 = R_7 = H; R_4 = NHAc$
22: $R_1, R_2 = H, OH; R_3 = R_5 = R_6 = R_7 = H; R_4 = NHAc$

Chart I

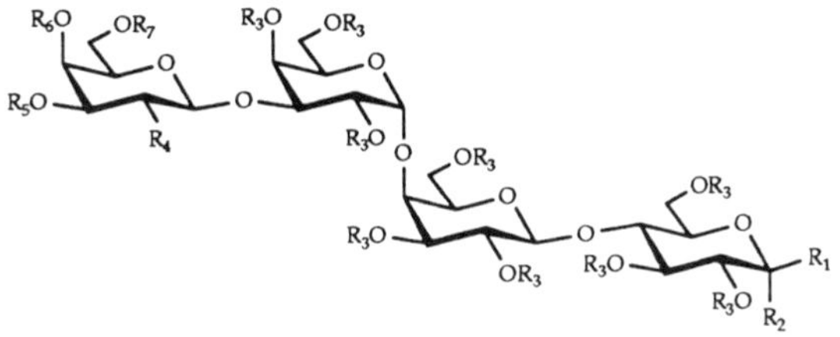

23: R_1 = O-TMSEt; R_2 = H; R_3 = Bzl; R_4 = NPhth; R_5 = R_6 = R_7 = Ac
24: R_1 = O-TMSEt; R_2 = R_5 = H; R_3 = Bzl; R_4 = NHAc; R_6 R_7 = PhCH
25: R_1 = O-TMSEt; R_2 = H; R_3 = R_5 = R_6 = R_7 = Ac; R_4 = NHAc
26: R_1 = O-TMSEt; R_2 = R_3 = R_5 = R_6 = R_7 = H; R_4 = NHAc
27: R_1,R_2 = H,OH; R_3 = R_5 = R_6 = R_7 = H; R_4 = NHAc

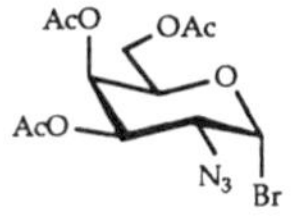

28

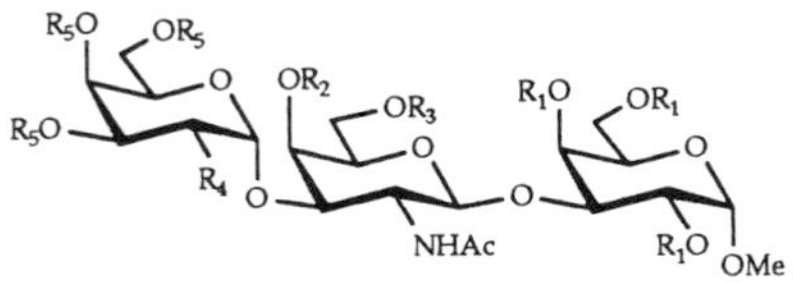

29: R_1 = Bzl; R_2,R_3 = PhCH; R_4 = N_3; R_5 = Ac
30: R_1 = R_3 = R_4 = R_5 = Ac; R_2 = NHAc
31: R_1 = R_3 = R_4 = R_5 = H; R_2 = NHAc

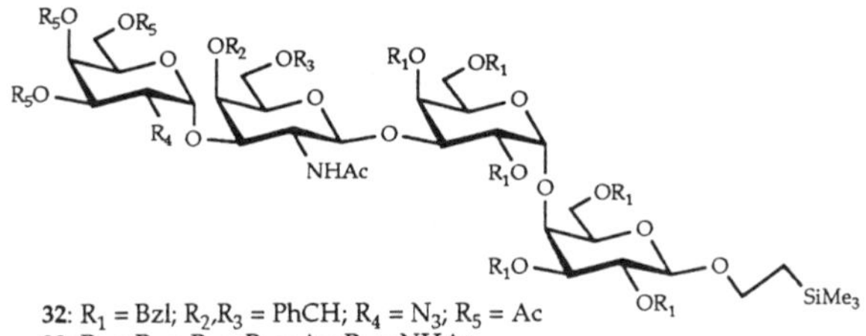

32: R_1 = Bzl; R_2,R_3 = PhCH; R_4 = N_3; R_5 = Ac
33: R_1 = R_2 = R_3 = R_5 = Ac; R_4 = NHAc
34: R_1 = R_2 = R_3 = R_5 = H; R_4 = NHAc

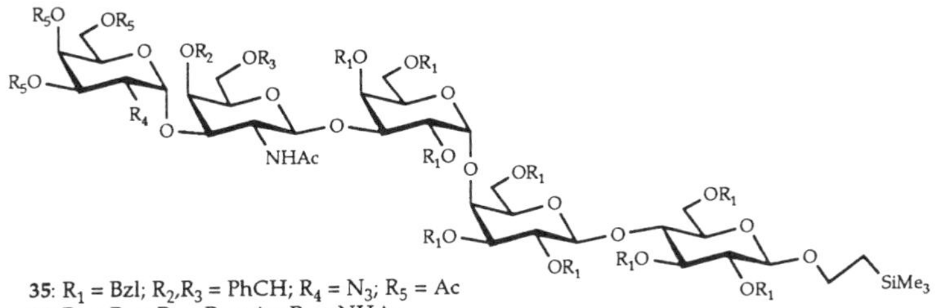

35: R_1 = Bzl; R_2,R_3 = PhCH; R_4 = N_3; R_5 = Ac
36: R_1 = R_2 = R_3 = R_5 = Ac; R_4 = NHAc
37: R_1 = R_2 = R_3 = R_5 = H; R_4 = NHAc

Chart II

Saccharides **16**, **19**, and **24** were glycosylated with the azidodeoxy bromide **28** (*52*), using silver triflate as promoter, to give the tri-, tetra-, and pentasaccharide glycosides **29** (41% + 15% β glycoside), **32** (69%), and **35** (74%), respectively. As expected (*53*), the less reactive tri- and tetrasaccharide acceptors **19** and **24** permitted selective α-glycosylation, whereas the more reactive disaccharide **16** was less selective.

The azido groups in compounds **29**, **32**, and **35** were reduced with nickel boride (NaBH₄/NiCl₂/H₃BO₃) (*54*) and the products were deblocked in the usual way, acetylated and purified to give **30**, **33**, and **36**. Deacetylation gave the tri-, tetra-, and pentasaccharide glycosides **31** (71%), **34** (54%), and **37** (51%).

Transformation of TMSEt Glycosides into DIB-related Glycosides. Apart from the efficient anomeric deblocking of TMSEt glycosides, as exemplified by the synthesis of **22** and **27** above, TMSEt glycosides undergo a highly stereoselective transformation into β-1-*O*-acyl sugars (*13a*). The TMSEt glycoside **38** (obtained in 90% yield by hydrogenolysis of **18** followed by hydrazinolysis and acetylation) was treated with acetic anhydride/boron trifluoride etherate (*13a*) to give the acetate **39** (89%; β/α 3:1). Glycosylation of 3-bromo-(2-bromomethyl)propanol (DIBOL) (*27*) with **39**, using boron trifluoride etherate as promoter (*55*), gave the DIB glycoside **40** (46% + 0.3% α-glycoside) shown in Chart III.

Reduction of the dibromide **40** with tributyltin hydride/azo-bis-isobutyronitrile gave isobutyl glycoside **41** (65%), which was deacetylated to give **42** (93%). Similar reductions were earlier performed by hydrogenolysis (*14*), which gave low yield and several byproducts with **40**.

Finally, cesium carbonate-mediated nucleophilic substitution of bromide by alkyl thiols is an efficient process that permits the preparation of neoglycolipids with varying alkyl chain lengths (*14*). Treatment of **40** with butyl thiol gave the bis-sulfide glycoside **43** (84%), which on oxidation (*14*) with m-chloroperbenzoic acid gave the corresponding bis-sulfone **44** (89%). Deacetylation of **44** gave the bis-sulfone glycoside **45**, which is potentially useful as inhibitor of bacterial adhesion (*10*).

Conformational Analysis

Conformational analysis of saccharides belonging to the globo series of glycolipids has been performed by several research groups, using a combination of X-ray crystallography, NMR spectroscopy and computational methods. Only a few crystal structures have been determined, whereas NMR studies have yielded preferred conformations for several of the oligosaccharide fragments of the Forssman pentasaccharide. We plan to use the complete set of di– pentasaccharides to investigate the validity of an NMR/computer calculation-based additivity approach to conformational analysis of larger saccharides.

Crystal and NMR data are being used as starting values in computations of

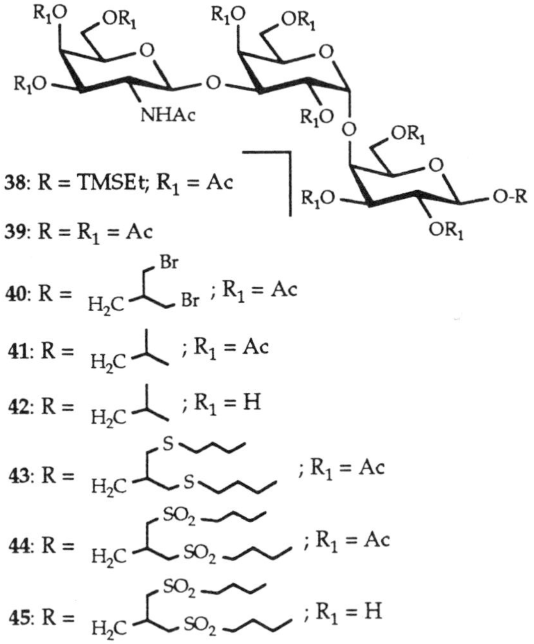

38: R = TMSEt; R_1 = Ac

39: R = R_1 = Ac

40: R = H_2C (Br, Br) ; R_1 = Ac

41: R = H_2C ; R_1 = Ac

42: R = H_2C ; R_1 = H

43: R = H_2C (S, S) ; R_1 = Ac

44: R = H_2C (SO_2, SO_2) ; R_1 = Ac

45: R = H_2C (SO_2, SO_2) ; R_1 = H

Chart III

saccharide conformations. However, the crystal conformation can be quite different from that in solution (see below). NMR analyses are usually based on NOE experiments, which give distance constraints to be added to the computational force fields, thereby aiding in the search for realistic conformations.

Oxygen-induced deshielding of carbohydrate ring hydrogens (56, 57) is a useful probe for conformational analysis. A strong (≥0.4 ppm) deshielding indicates a preferred conformation, with oxygen in close contact with the hydrogen atom. We consider these deshielding effects to be at least as reliable as NOE measurements for the determination of preferred conformations of oligosaccharides.

Galabiose (fragment C in Figure 1) is a particularly good example, where H-5 of the α-D-Gal unit is shifted downfield due to a close contact with O-3 of the β-D-Gal unit. Conformations that satisfy a van der Waals contact [(C)-H···O distance ≤2.7 Å] were calculated by the HSEA force field (57) and the Φ/Ψ-angles -39°/-15° were found for the conformation with the lowest energy. These values (corroborated by NOE-measurements) (57) are quite different from those found in the galabiose crystal (11) (-18°/+35°); thus, crystal data should be used with caution in conformational calculations of saccharides.

We have determined the chemical shift of H-5 in the α-D-Gal unit of the Forssman fragments A-I, (Table III) as well as in other galabiose derivatives. In all cases where an intact galabiose residue is present, the H-5 chemical shift is close to 4.35 ppm, representing a down-field shift of 0.4-0.5 ppm as compared to methyl α-D-galactopyranoside or methyl 3-deoxy-β-D-galabioside. This consistency in the H-5 chemical shift suggests that the conformation is highly conserved in the galabiose part of the various saccharides of the Forssman antigen.

The Φ/Ψ angles of the disaccharide fragments of the Forssman pentasaccharide were calculated using the dihedral driver option of the MacMimic/MM2(87) package (58). The Φ/Ψ-values for the galabiose residue mentioned above and the values for low-energy conformers of the remaining three disaccharide fragments were used as starting values for energy minimization of the Forssman pentasaccharide. Several conformers with small Φ/Ψ differences and with energies within 0.2 kcal/mol were obtained. The Φ/Ψ values for one of these conformers are shown in Table IV together with published low-energy conformations of the corresponding globotetraosylceramide (cf. fragment H in Figure 1). A stereo view and a space-filling model of the TMSEt glycoside of the Forssman pentasaccharide (37) is depicted in Figure 4, clearly showing the L-shape of this compound. It is the galabiose residue that makes up the convex-bend part of the molecule, thereby making it easy for galabiose-binding proteins to get access to the binding surface of the saccharide.

Table III. Chemical Shifts of the H-5 Hydrogen Atom of the α-D-Gal Unit of Various Glycosides Related to the Forssman Pentasaccharide

Saccharide	Chemical shift (ppm)
GalNAcα1-3GalNAcβ1-3Galα1-4Galβ1-4GlcβOTMSEt (**37**)	4.37
GalNAcβ1-3Galα1-4Galβ1-4GlcβOTMSEt (**26**)	4.33
GalNAcβ1-3Galα1-4Galβ1-4GlcOH (**27**)	4.31
Galα1-4Galβ1-4GlcβOTMSEt (*10*)	4.31
GalNAcα1-3GalNAcβ1-3Galα1-4GalβOTMSEt (**34**)	4.36
GalNAcβ1-3Galα1-4GalβOTMSEt (**21**)	4.36
GalNAcβ1-3Galα1-4GalβOCH₂CH(CH₂SO₂C₄H₉)₂ (**45**)	4.37
GalNAcβ1-3Galα1-4GalOH (**22**)	4.36
Galα1-4GalβOTMSEt (*13a*)	4.34
Galα1-4GalβOiBu (*14*)	4.37
Galα1-4GalβOMe (*28*)	4.34
Galα1-4Gal(6-deoxy)βOMe (*30*)	4.39
Galα1-4Gal(3-deoxy)βOMe (*29*)	3.97
GalNAcα1-3GalNAcβ1-3GalαOMe (**31**)	~3.79
GalNAcβ1-3GalαOMe (**17**)	~3.75
GalαOMe	3.78

Table IV. Φ/Ψ Angles for the Forssman Pentasaccharide and Globotetraosylceramide

Compound	Method	Φ/Ψ^a (°)			
		GalNAcα	GalNAcβ	Galα	Galβ
Gb₅TMSEt (**37**)	MM2(87)	-32/-32	+24/-59	-40/-14	+47/-12
Gb₄Cer (*59*)	MM2(85)		+35/-57	-37/-16	+66/-7
Gb₄Cer (*60*)	HSEA		+40/-50	-50/-10	+20/-10
Gb₄Cer (*9*)	HSEA		+57/-7	-39/-11	+53/+3

^{a}Φ: H-1'—C-1'—O-1'—C-3/4; Ψ: C-1'—O-1'—C-3/4—H3/4.

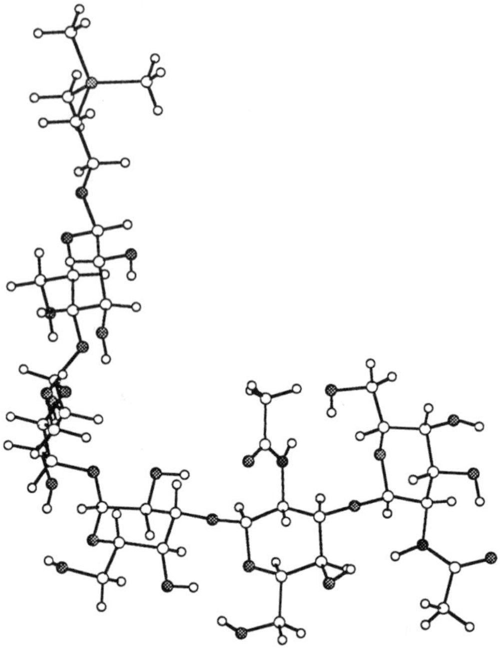

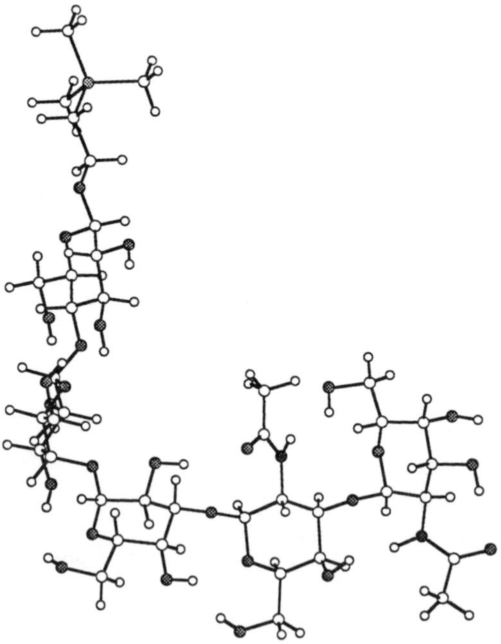

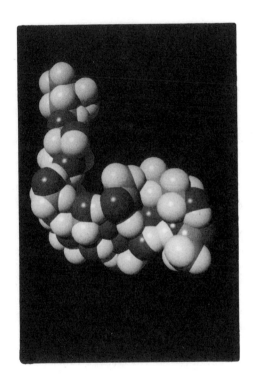

Figure 4. Stereo view and space-filling model of the TMSEt glycoside of the Forssman pentasaccharide (**37**).

Acknowledgments. This work was supported by grants from The Swedish Natural Science Research Council, The Swedish National Board for Technical Development, and Symbicom AB, Lund, Sweden.

Literature cited

1. Normark, S.; Båga, M.; Göransson, M.; Lindberg, F. P.; Lund, B.; Norgren, M.; Uhlin, B.-E. In *Microbial Lectins and Agglutinins*; Mirelman, D., Ed., Wiley Series in Ecological and Applied Microbiology; John Wiley & Sons, New York, 1986, 113-143.

2. Lingwood, C. A.; Law, H.; Richardson, S.; Petric, M.; Brunton, J. L.; De Grandis, S.; Karmali. M. *J. Biol. Chem.* **1987**, *262*, 8834.

3. (a) Jacewicz, M.; Clausen, H.; Nudelman. E.; Donohue-Rolfe, A.; Keusch, G. T. *J. Exp. Med.* **1986**, *163*, 1391. (b) Lindberg, A. A.; Brown, J. E.; Strömberg, N.; Westling-Ryd, M.; Schultz, J. E.; Karlsson, K. A. *J. Biol. Chem.* **1987**, *262*, 1779.

4. Nudelman, E.; Kannagi, R.; Hakomori, S.; Parsons, M.; Lipinski,M.; Wiels, J.; Fellous, M.; Tursz, T. *Science*, **1983**, *220*, 509.

5. Sweely, C. C.; Klionski, B. *J. Biol. Chem.* **1963**, *238*, 3148.

6. Hultgren, S. J.; Lindberg, F.; Magnusson, G.; Kihlberg, J.; Tennent, J. M.; Normark, S. *Proc. Natl. Acad. Sci. USA*, **1989**, *86*, 4357.

7. Holmgren, A.; Brändén, C.-I. *Nature*, **1989**, *342*, 248.

8. (a) Källenius, G.; Möllby, R.; Svenson, S. B.; Winberg, J. Lundblad, A.; Svensson, S.; Cedergren, B. *FEMS Microbiol. Lett.* **1980**, *7*, 297. (b) Leffler, H.; Svanborg-Edén, C. *FEMS Microbiol. Lett.* **1980**, *8*, 127.

9. Bock, K.; Breimer, M. E.; Brignole, A.; Hansson, G. C.; Karlsson, K.-A.; Larsson, G.; Leffler, H.; Samuelsson, B. E.; Strömberg, N.; Svanborg-Edén, C.; Thurin, J. *J. Biol. Chem.* **1985**, *260*, 8545.

10. Kihlberg, J.; Hultgren, S. J.; Normark, S.; Magnusson, G. *J. Am. Chem. Soc.* **1989**, *111*, 6364.

11. Svensson, G.; Albertsson, J.; Svensson, C.; Magnusson, G.; Dahmén, J. *Carbohydr. Res.* **1986**, *146*, 29.

12. Strömberg, N.; Marklund, B.-I; Lund, B.; Ilver, D.; Hamers, A.; Gaastra, W.; Karlsson, K.-A.; Normark, S. *EMBO J.* **1990**, *9*, 2001.

13. (a) Jansson, K.; Ahlfors, S.; Frejd, T.; Kihlberg, J.; Magnusson, G.; Dahmén, J.; Noori, G.; Stenvall, K. *J. Org. Chem.* **1988**, *53*, 5629. (b) Jansson, K.; Noori, G.; Magnusson, G. *J. Org. Chem.* **1990**, *55*, 3181.

14. Magnusson, G.; Ahlfors, S.; Dahmén, J.; Jansson, K.; Nilsson, U.; Noori, G.; Stenvall, K.; Tjörnebo, A. *J. Org. Chem.* **1990**, *55*, 3932.

15. Paulsen, H.; Bünsch, A. *Carbohydr. Res.* **1982**, *100*, 143.

16. Jones, J. K. N. ; Reid, W. W. *J. Chem Soc.* **1955**, 1890.

17. Aspinall, G. O.; Fanshawe, R. S. *J. Chem Soc.* **1961**, 4215.

18. Chacon-Fuertes, M. E.; Martin-Lomas, M. *Carbohydr. Res.* **1975**, *43*, 51.

19. Gent, P. A.; Gigg, R.; Penglis, A. A. *J. Chem. Soc., Perkin Trans. 1*, **1976**, 1395.

20. Cox, D. D.; Metzner, E. K.; Reist, E. J. *Carbohydr. Res.* **1978**, *62*, 245.
21. Milat, M.-L.; Zollo, P. A.; Sinaÿ, P.; *Carbohydr. Res.* **1982**, *100*, 263.
22. Garegg, P. J.; Hultberg, H. *Carbohydr. Res.* **1982**, *110*, 261.
23. Dahmén, J.; Frejd, T.; Lave, T.; Lindh, F.; Magnusson, G.; Noori, G.; Pålsson, K. *Carbohydr. Res.* **1983**, *113*, 219.
24. Dahmén, J.; Frejd, T.; Grönberg, G.; Lave, T.; Magnusson, G.; Noori, G. *Carbohydr. Res.* **1983**, *116*, 303.
25. Dahmén, J.; Frejd, T.; Magnusson, G.; Noori, G. *Carbohydr. Res.* **1982**, *111*, c1.
26. Dahmén, J.; Frejd, T.; Grönberg, G.; Lave, T.; Magnusson, G.; Noori, G. *Carbohydr. Res.* **1983**, *118*, 292.
27. Ansari, A. A.; Frejd, T.; Magnusson, G. *Carbohydr. Res.* **1987**, *161*, 225.
28. Garegg, P. J.; Oscarsson, S. *Carbohydr. Res.* **1985**, *137*, 270.
29. Kihlberg, J.; Frejd, T.; Jansson, K.; Magnusson, G. *Carbohydr. Res.* **1986**, *152*, 113.
30. Kihlberg, J.; Frejd, T.; Jansson, K.; Sundin, A.; Magnusson, G. *Carbohydr. Res.* **1988**, *176*, 271
31. Kihlberg, J.; Frejd, T.; Jansson, K.; Magnusson, G. *Carbohydr. Res.* **1988**, *176*, 287
32. Kihlberg, J.; Frejd, T.; Jansson, K.; Kitzing, S.; Magnusson, G. *Carbohydr. Res.* **1989**, *185*, 171.
33. Cox, D. D.; Metzner, E. K.; Reist, E. J. *Carbohydr. Res.* **1978**, *63*, 139.
34. Dahmén, J.; Frejd, T.; Magnusson, G.; Noori, G.; Carlström, A.-S. *Carbohydr. Res.* **1984**, *127*, 15.
35. Jacquinet, J.-C.; Sinaÿ, P. *Carbohydr. Res.* **1985**, *143*, 143.
36. Koike, K.; Sugimoto, M.; Sato, S.; Ito, Y.; Nakahara, Y.; Ogawa, T. *Carbohydr. Res.* **1987**, *163*, 189.
37. Shapiro, D.; Acher, A. *J. Chem. Phys. Lipids* **1978**, *22*, 197.
38. Nicolaou, K. C.; Caulfield, T.; Katoaka, H.; Kumazawa, T. *J. Am. Chem. Soc.* **1988**, *110*, 7910.
39. Nicolaou, K. C.; Caulfield, T.; Katoaka, H. *Carbohydr. Res.* **1990**, *202*, 177.
40. Paulsen, H.; Bünsch, A. *Carbohydr. Res.* **1982**, *101*, 23.
41. Leontein, K.; Nilsson, M.; Norberg, T. *Carbohydr. Res.* **1985**, *144*, 231.
42. Paulsen, H.; Bünsch, A. *Liebigs Ann. Chem.* **1981**, 2204.
43. Paulsen, H.; Bünsch, A. *Carbohydr. Res.* **1982**, 100, 143.
44. Nunomura, S.; Ogawa, T. *Tetrahedron Lett.* **1988**, *29*, 5681.
45. Kameyama, A.; Ishida, H.; Kiso, M.; Hasegawa, A. *J. Carbohydr. Chem.* **1991**, *10*, 549.
46. David, S.; Hanessian, S. *Tetrahedron,* **1985**, *41*, 643.
47. Ogawa, T.; Yamamoto, H. *Agric. Biol. Chem.* **1985**, *49*, 475.
48. Iversen, T.; Bundle, D. R. *Carbohydr. Res.* **1982**, *103*, 29.
49. Garegg, P. J.; Hultberg, H. *Carbohydr. Res.* **1981**, *93*, c10.
50. Kong, F.; Lu, D.; Zhou, S. *Carbohydr. Res.* **1990**, *198*, 141.

51. Nilsson, U.; Ray, A. K.; Magnusson, G. *Carbohydr. Res.* **1990**, *208*, 260.
52. Lemieux, R. U.; Ratcliffe, R. M. *Can. J. Chem.* **1979**, *57*, 1244.
53. Paulsen, H. *Angew. Chem. Int. Ed. Engl.* **1982**, *21*, 155.
54. Paulsen, H.; Sinnwell, V. *Chem. Ber.* **1978**, *111*, 879.
55. Magnusson, G.; Noori, G.; Dahmén, J.; Frejd, T.; Lave, T. *Acta Chem. Scand.* **1981**, *B35*, 213.
56. Thögersen, H.; Lemieux, R. U.; Bock, K.; Meyer, B. *Can. J. Chem.* **1982**, *60*, 44.
57. Bock, K.; Frejd, T.; Kihlberg, J.; Magnusson, G. *Carbohydr. Res.* **1988**, *176*, 253.
58. InStar Software, Ideon Research Park, S-223 70 Lund, Sweden.
59. Poppe, L.; von der Lieth, C.-W.; Dabrowski, J. *J. Am. Chem. Soc.* **1990**, *112*, 7762.
60. Scarsdale, J. N.; Yu, R. K.; Prestegard, J. H. *J. Am. Chem. Soc.* **1986**, *108*, 6778.

RECEIVED March 25, 1992

Chapter 9

Synthesis and Immunochemistry of Carbohydrate Antigens of the β-Hemolytic *Streptococcus* Group A

B. Mario Pinto

Department of Chemistry, Simon Fraser University, Burnaby, British Columbia V5A 1S6, Canada

Progress towards the synthesis of increasingly complex oligosaccharides corresponding to the cell-wall polysaccharide of the β-hemolytic *Streptococcus* Group A is described. Strategies based on the preparation of a key branched trisaccharide unit, α-L-Rhap-(1-2)-[β-D-GlcpNAc-(1-3)]-α-L-Rhap, or a linear trisaccharide unit, β-D-GlcpNAc-(1-3)-α-L-Rhap-(1-3)-α-L-Rhap, each of which function as both a glycosyl acceptor and donor, have been pursued. Disaccharide, linear trisaccharide, and branched tri-, tetra-, penta- and hexasaccharides have been obtained. Furthermore, a convergent synthetic route, based on a fully functionalized branched trisaccharide block, has been developed. This route has potential for the elaboration of even higher-order structures. The compounds have been obtained as their propyl and/or 8-methoxycarbonyloctyl glycosides. The latter compounds have been coupled to bovine serum albumin or horse hemoglobin to yield the corresponding glycoconjugates. Immunochemical studies employing the glycoconjugates and the panel of oligosaccharide haptens have served to characterize rabbit polyclonal and mouse monoclonal antibodies raised against the glycoconjugates or a killed bacterial vaccine, respectively. The branch point of the *Streptococcus* Group A antigen appears to be a crucial element of the epitope recognized by both polyclonal and monoclonal antibodies that are able to bind the native antigen. An IgM monoclonal antibody that recognizes an extended binding site has been identified as a suitable candidate for the design of immunodiagnostic reagents.

The β-hemolytic *Streptococcus* Group A is one of the primary infective agents in humans, causing streptococcal pharyngitis (strep throat)

0097–6156/93/0519–0111$06.25/0

(*1,2*). In certain cases, the streptococcal infection acts as a trigger of acute rheumatic fever (*2*). Rheumatic fever can cause immediate fatal damage to the heart muscle or it can lead to long-term lethal effects by scarring heart valves (*2*). Evidence to date has led to the hypothesis that an abnormal immune response to *Streptococcus* Group A infection results in an autoimmune reaction operating in rheumatic fever in which cells directed against streptococcal antigens attack host tissue antigens, particularly in the heart (*1-7*). Thus, Goldstein *et al.* (*3,4*) have demonstrated immunological cross reaction between *Streptococcus* Group A polysaccharide and the structural glycoproteins of bovine and human heart valves. In addition, adults with chronic rheumatic valvular disease and children with persistent valvular disease showed elevated levels of antibody to the streptococcal Group A carbohydrate (*5*), and antibodies in the sera of rheumatic fever patients were found to be cross-reactive with extracts of heart valves (*3*). Repeated and prolonged immunization of rabbits with *Streptococcus* Group A polysaccharide or peptidoglycan has also led to the appearance of cardiac lesions and circulating antibodies (*6*). Immunopathology with labelled antibodies subsequently indicated binding to the damaged areas and also the binding of anti-*Strep*-A antibodies to the connective tissue of the cardiac valves and coronary vessels (*6*). However, the exact mechanism by which the streptococcal infection is linked to the onset and propagation of rheumatic heart disease remains to be resolved (*1,7*).

Several years ago, we proposed to develop and characterize monoclonal antibody-based immunodiagnostic reagents with specificity for various carbohydrate epitopes present on the cell wall polysaccharide of *Streptococcus* Group A with which to probe the relationship between *Streptococcus* Group A infections and rheumatic heart disease.

As part of this program, we have focused in recent years on the synthesis of increasingly complex oligosaccharides that represent different epitopes of the *Streptococcus* Group A cell wall polysaccharide, and on the development of convergent synthetic routes that would readily furnish higher-order structures. We describe herein our progress in this area and also present the immunochemical characterization of polyclonal and monoclonal antibodies by use of the oligosaccharide haptens and glycoconjugates derived therefrom.

Synthesis of Oligosaccharide Haptens and Glycoconjugates

The cell-wall carbohydrates of the Group A *Streptococcus* are comprised of a rhamnose backbone consisting of alternating α-L-(1-2) and α-L-(1-3) linkages, with β-D-N-acetylglucosamine residues attached to the 3-positions of the rhamnose backbone (*8,9*).

$$
\left[
\begin{array}{cccc}
\mathbf{A'} & \mathbf{B'} & \mathbf{A} & \mathbf{B} \\
\!\!\!-\alpha\text{-L-Rha}p\text{-}(1\text{-}2)\text{-}\alpha\text{-L-Rha}p\text{-}(1\text{-}3)\text{-}\alpha\text{-L-Rha}p\text{-}(1\text{-}2)\text{-}\alpha\text{-L-Rha}p\text{-}(1\text{-}3)\!- \\
|3 |3 \\
|1 |1 \\
\text{-}\beta\text{-D-Glc}p\text{NAc} \text{-}\beta\text{-D-Glc}p\text{NAc} \\
\mathbf{C'} \mathbf{C}
\end{array}
\right]_n
$$

Retrosynthetic analysis indicated that disconnections based on key linear (AB'C') or branched B(C)A trisaccharide sequences or an α-L-Rha*p*-(1-2)-α-L-Rha*p*-(1-3)-α-L-Rha*p*-(1-2)-α-L-Rha*p*-(1-3)-backbone might be desirable. We report here our synthetic efforts based on the former two strategies and the use of König-Knorr glycosylation reactions.

We envisaged that a trisaccharide unit such as 1 would be ideal for the block synthesis of higher order structures (Scheme 1). The salient features of this molecule are the presence of different ester functions at C-2 and C-2', benzyl ethers and an N-phthalimido function as persistent blocking groups on the glucosamine ring, and an allyl glycoside. In principle, compound 1 would readily afford a glycosyl donor 2 by successive 1) isomerization of the allyl glycoside to the prop-1-enyl glycosides (*10,11*), 2) hydrolysis to give the hemiacetals (*12*), and 3) formation of the glycosyl halides by use of Vilsmeier-Haack reagents (*13-15*). Alternatively, selective removal of the 2'-*O*-acetyl group would give a trisaccharide acceptor 3 (Scheme 1).

The proposed synthesis of the key trisaccharide 1 derives from the glycosylation of the acceptor 4 (*16*) with the disaccharide halides 5 (Scheme 2). However, despite numerous attempts, using various combinations of the promoters, bases, and reaction medium, we have been unable to achieve a satisfactory outcome of this reaction (Tixidre, A.; Pinto, B. M., unpublished results). Rather, the reactions have proceeded with poor stereoselectivity and have, surprisingly, yielded a preponderance of the unwanted β-anomer. A further complication was the formation of a significant amount of the 1,2 elimination product 7 from the donors 5 (Scheme 2). In contrast, the corresponding disaccharide chloride in which the substituent at C-2 is a benzoate ester and not an acetate reacts with 4 under silver trifluoromethanesulfonate (AgTfl) promotion in the presence of 1,1,3,3-tetramethylurea (TMU) (*17*) to give exclusively the α-anomer

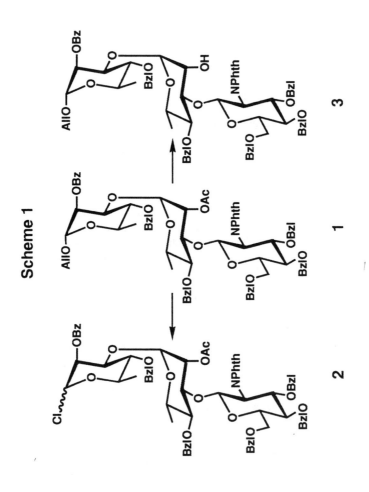

Scheme 1

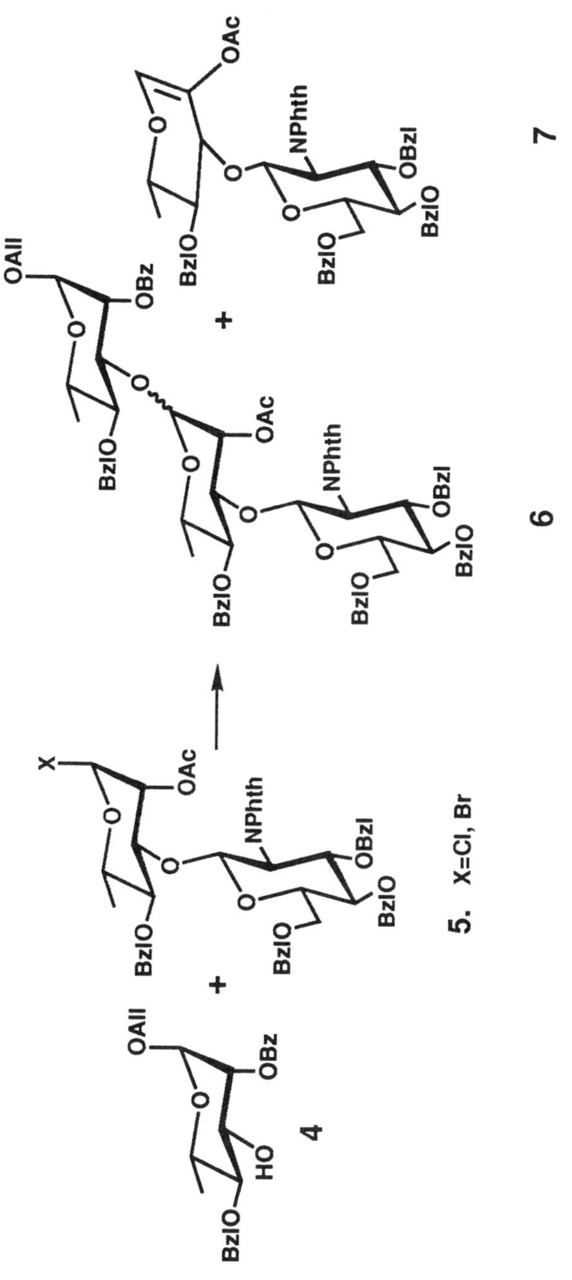

Scheme 2

in good yield (Tixidre, A.; Pinto, B. M., unpublished results). Analogously, the reaction of **4** with the corresponding disaccharide chloride having acetate esters on the glucosamine residue and a benzoate ester at C-2 proceeds without incident to give the α anomer (*18*).

It appeared, therefore, that the presence of a benzoate ester at C-2 of a disaccharide donor such as **5** was essential for the desired stereochemical outcome of the reaction. That this was indeed the case was further indicated by two other observations. Firstly, glycosylation of **4** (AgTfl/TMU) with the disaccharide chloride **8** (containing benzoate esters on the glucosamine unit and an acetate ester at C-2) gave a 6:1 β:α mixture of trisaccharides **9** (Scheme 3), albeit in low yield (36%) (*19*). Secondly, the disaccharide chloride **10** [containing 2-(trimethylsilyl)ethoxymethyl (SEM) acetals as protecting groups on the glucosamine residue and a benzoate ester at C-2] reacted (AgTfl/TMU) with the acceptor **11** to give the trisaccharide **12** in 81% yield (Scheme 4) (*20*).

The trisaccharide **12** appeared to be a reasonable candidate as a key intermediate for the elaboration of higher order structures. However, glycosylation at C-2' of **13** (*21*) now posed a serious problem. Branching at this position with a variety of donors, using different combinations of promoters, bases, and solvents, was totally unsatisfactory. The best result (50%) was obtained for the monosaccharide bromide **14** under promotion with mercury (II) cyanide/mercury (II) bromide (Scheme 5) (*20*). This reaction did afford the tetrasaccharide **15** but the overall scheme did not appear promising as an efficient route to higher order oligosaccharides.

Although a convergent synthetic route based on a linear trisaccharide unit was not realized, the various synthetic attempts did provide some key disaccharides for the synthesis of more complex oligosaccharides that represent different epitopes. Thus, the disaccharide **16** was glycosylated with **17** to give the branched trisaccharide **18**, as its allyl glycoside, and an analogous reaction with the corresponding 8-methoxycarbonyloctyl glycoside **19** afforded the trisaccharide **20** (Scheme 6) (*22*). Similarly, reaction of **19** with the trisaccharide donor **21** gave the branched pentasaccharide **22**, as its 8-methoxycarbonyloctyl glycoside (Scheme 7) (*22*). The corresponding glycosylation of the allyl glycoside **23** yielded the branched pentasaccharide **24** (Scheme 8) (*19*).

A convergent synthetic route based on the branched trisaccharide unit was pursued next (Reimer, K. B.; Harris, S. L.; Pinto, B. M.; Varma, V. *Carbohydr. Res.*, in press). Accordingly,

Scheme 3

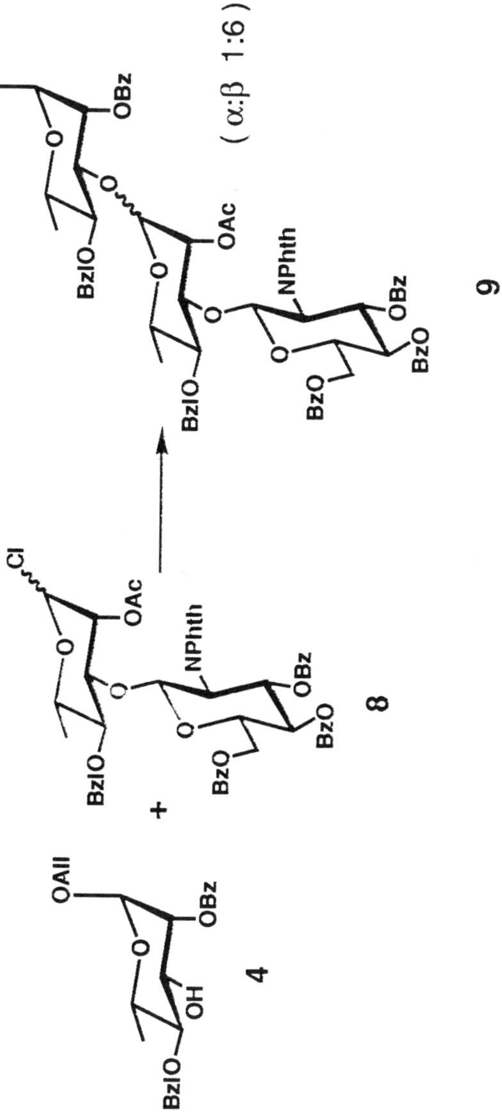
BzlO
OAll
OH
OBz
4
+
BzlO
Cl
OAc
NPhth
BzO
OBz
BzO
8
BzlO
OAll
OBz
BzlO
OAc
NPhth
BzO
OBz
BzO
9
(α:β 1:6)

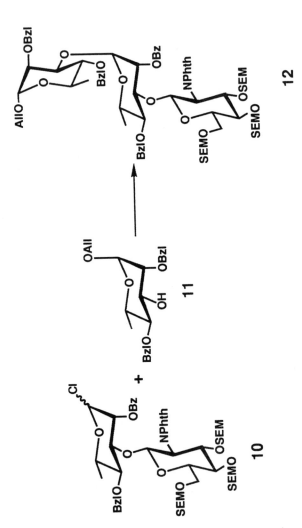

Scheme 4

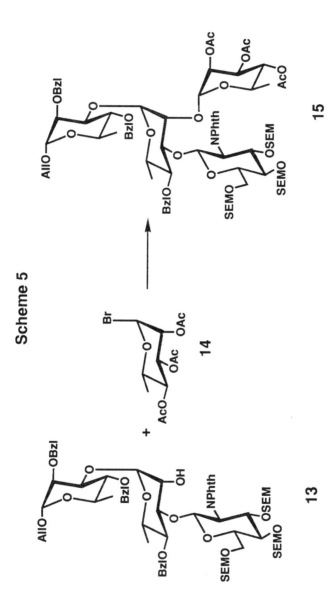

Scheme 5

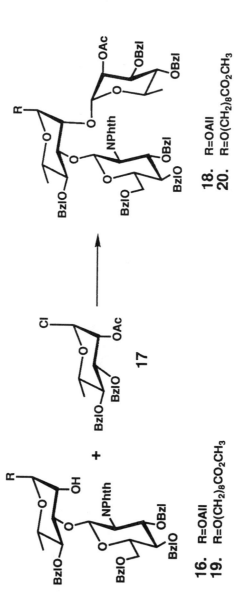

Scheme 6

Scheme 7

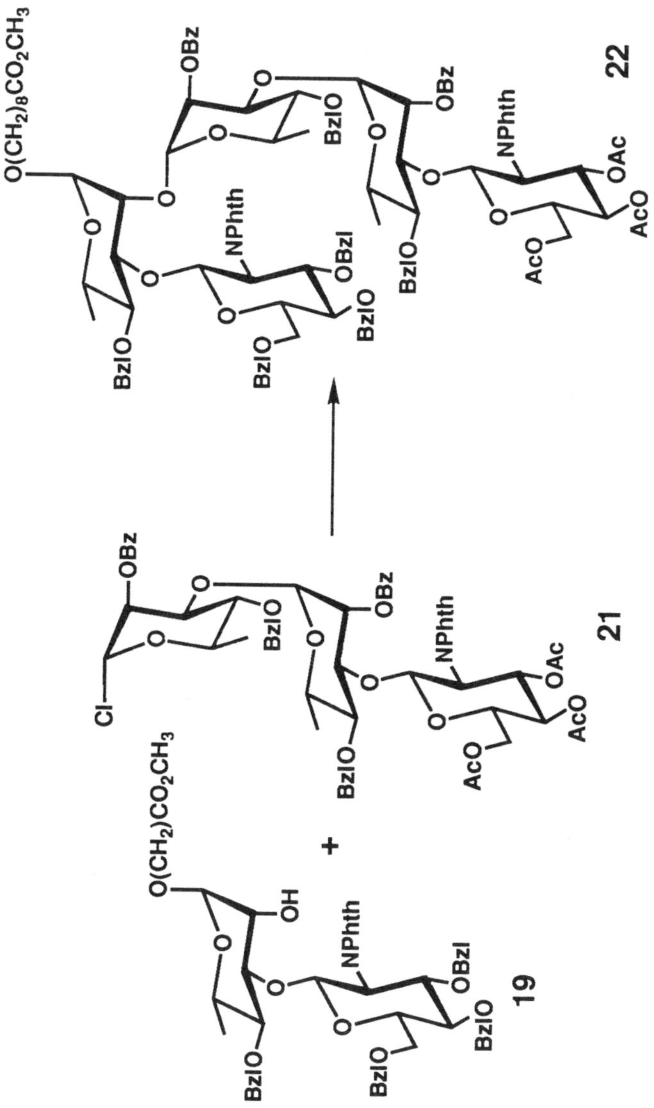
BzlO
OH
O(CH2)CO2CH3
NPhth
OBzl
19
+
Cl
OBz
OAc
AcO
21
O(CH2)8CO2CH3
22

Scheme 8

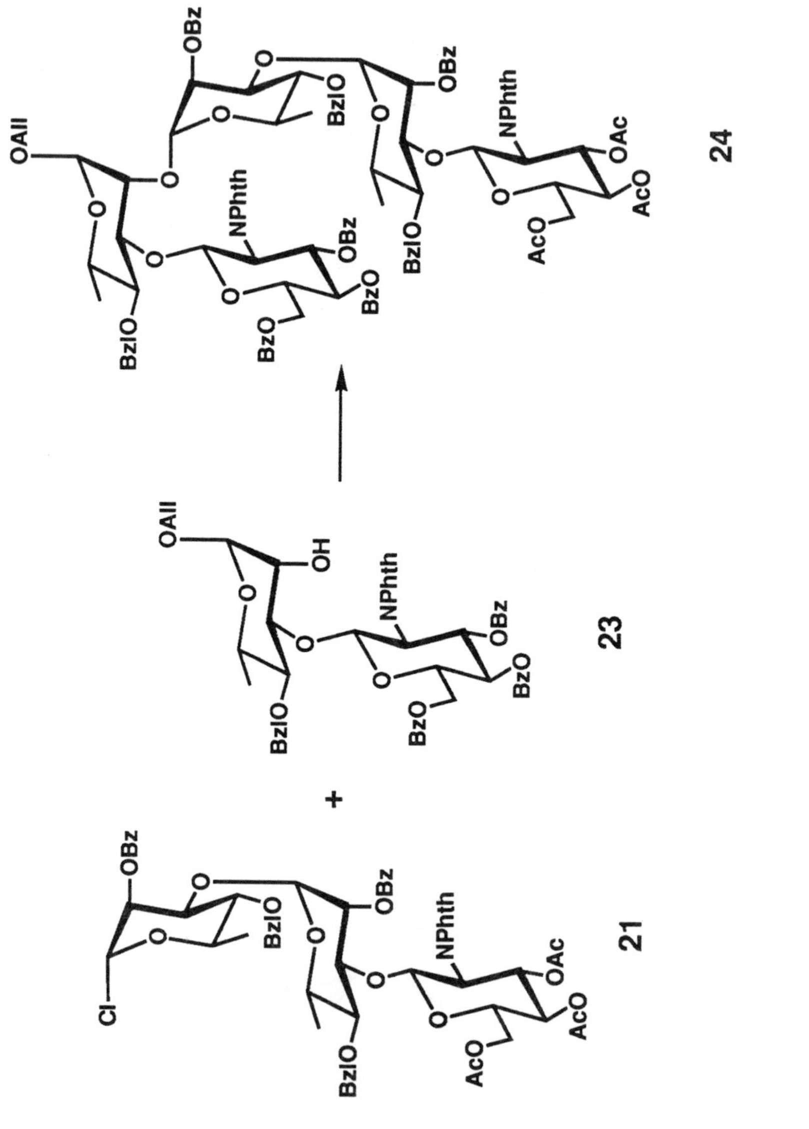
Cl
OBz
BzlO
OBz
NPhth
OAc
AcO
AcO
BzlO
21
+
OAll
OH
NPhth
OBz
BzO
BzO
BzlO
23
OAll
BzlO
BzO
BzO
NPhth
OBz
OBz
BzlO
OBz
NPhth
OAc
AcO
AcO
BzlO
24

the disaccharide acceptor **23** was reacted with the donor **25** (*23*) to give the fully functional branched trisaccharide **26** (Scheme 9). Compound **26** was capable of functioning adequately as both a donor and acceptor for the synthesis of higher order structures. Thus, selective removal of the SEM acetal in 3% methanolic HCl (*23*) afforded the acceptor **27**, and deallylation and conversion to the chloride, as before (*23*), gave the donor **28** (Scheme 9). Glycosylation of **27** with **28** (AgTfl/TMU) yielded the hexasaccharide **29** with the desired α stereochemistry in 50% yield (Scheme 10). The stereochemical outcome in the absence of a participating group at C-2 is significant but is not without precedent. For example, Bundle's group (*24-26*) have prepared several homopolymers of 1-2 α-linked perosamine (4,6-dideoxy-4-formamido-D-mannose) units. In these glycosylation reactions, there were other glycosyl residues at the 2-positions of the glycosyl donors. In addition, Ogawa *et al* have used both α-D-mannosyl bromides (*27*), and α-D-mannosyl trichloroacetimidates (*28*) to prepare oligosaccharides with α-stereoselectivity; in both cases the mannosyl donors had other sugar residues at the 2-positions. The stereochemical outcome of such reactions is not, however, guaranteed, as indicated by the results of Srivastava and Hindsgaul (*29*) who found extensive β-glycoside formation with mannosyl donors that were glycosylated at C-2.

The synthesis of the hexasaccharide **29**, by use of only three glycosylation reactions, defines an efficient, convergent synthetic route to higher order oligosaccharides of the *Streptococcus* Group A cell-wall polysaccharide. Compound **29** contains the same features as the parent trisaccharide, namely, a SEM group that can be selectively removed to give an acceptor and an allyl glycoside that can be manipulated to give a glycosyl donor. Therefore, in principle, a nonasaccharide or a dodecasaccharide could be derived from a fourth glycosylation reaction.

The haptens were obtained by successive treatment with 1) 3% methanolic HCl to remove the SEM group, 2) sodium methoxide in methanol to remove the benzoate esters, 3) hydrazinolysis of the phthalimido group and *N*-acetylation of the resultant amine, and 4) hydrogenolysis of the benzyl ethers and hydrogenation of the allyl group. The sequence furnished oligosaccharides as their n-propyl and 8-methoxycarbonyloctyl glycosides **30-38** (Figure 1). The latter were converted via hydrazide derivatives to acyl azide intermediates (*30-32*) which were coupled to the protein carriers, bovine serum albumin (BSA) and horse hemoglobin (HHb) to give the corresponding glycoconjugates (*22*). One of the hydrazides **39** was also used in subsequent immunochemical studies. The glycoconjugates **40-42** were used as antigens in immunochemical studies, as described below.

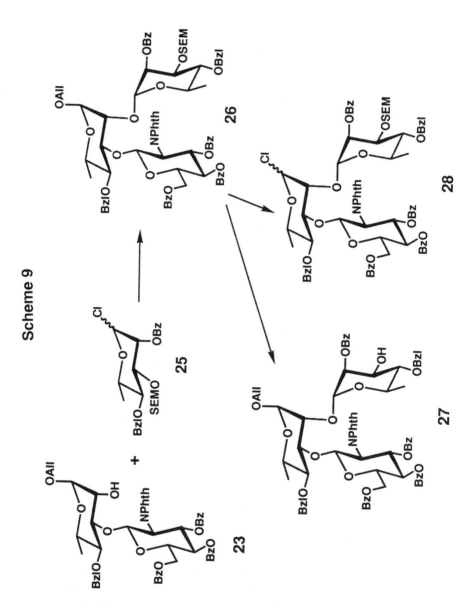

Scheme 9

Scheme 10

27 + 28

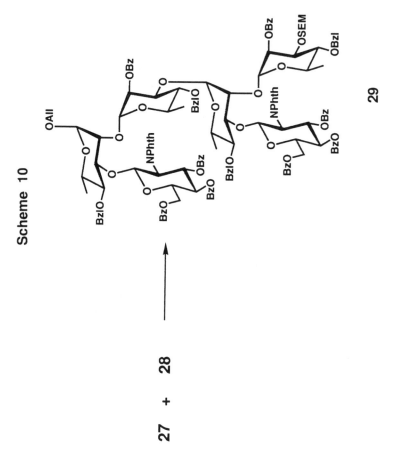

29

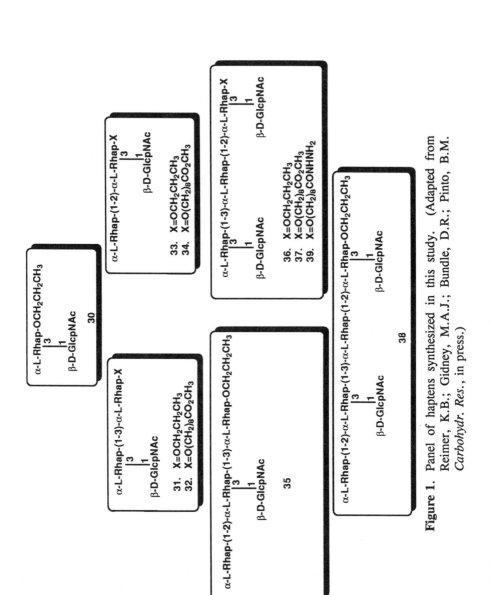

Figure 1. Panel of haptens synthesized in this study. (Adapted from Reimer, K.B.; Gidney, M.A.J.; Bundle, D.R.; Pinto, B.M. *Carbohydr. Res.*, in press.)

$$\left[\begin{array}{c} -\alpha\text{-L- Rha}p\ -(1\text{-}3)\text{-}\alpha\text{-L-Rha}p\text{-O(CH}_2)_8\text{CONH} \\ \overset{|3}{\underset{|1}{}} \\ \beta\text{-D-Glc}p\text{NAc} \end{array} \right]_n \text{---Protein}$$

Antigen **40a** : Protein = BSA
Antigen **40b** : Protein = HHb

$$\left[\begin{array}{c} -\alpha\text{-L-Rha}p\text{-}(1\text{-}2)\text{-}\alpha\text{-L- Rha}p\ \text{-O(CH}_2)_8\text{CONH} \\ \overset{|3}{\underset{|1}{}} \\ \beta\text{-D-Glc}p\text{NAc} \end{array} \right]_n \text{---Protein}$$

Antigen **41a** : Protein = BSA
Antigen **41b** : Protein = HHb

$$\left[\begin{array}{c} -\alpha\text{-L- Rha}p\text{-}(1\text{-}3)\text{-}\alpha\text{-L-Rha}p\text{-}(1\text{-}2)\text{-}\alpha\text{-L-Rha}p\ \text{-O(CH}_2)_8\text{CONH} \\ \overset{|3}{\underset{|1}{}} \qquad\qquad\qquad\qquad \overset{|3}{\underset{|1}{}} \\ \beta\text{-D-Glc}p\text{NAc} \qquad\qquad\qquad \beta\text{-Glc}p\text{NAc} \end{array} \right]_n \text{---Protein}$$

Antigen **42a** : Protein = BSA
Antigen **42b** : Protein = HHb

Immunochemical Studies. (Reimer, K. B.; Gidney, M. A. J.;
Bundle, D. R.; Pinto, B. M.
Carbohydr. Res., in press)

The antigens **40a**, **41a**, and **42a** were used to immunize three groups
of rabbits and the corresponding horse hemoglobin (HHb) conjugates
40b, **41b**, and **42b** were then used to assay carbohydrate-specific
antibody. Titration curves and a series of inhibition ELISA studies in
which disaccharide through pentasaccharide haptens were used as
inhibitors of antibody-glycoconjugate binding indicated that the
polyclonal antibodies raised against a particular glycoconjugate
exhibited moderate specificity for the oligosaccharide epitope of the
immunizing antigen. The antibodies generated by glycoconjugate **40a**
and **41a** cross-reacted with the branched pentasaccharide antigen **42b**.
This is not surprising since the pentasaccharide contains within its
structure both epitopes represented by the linear trisaccharide and
branched trisaccharide.

Only the antibodies raised to the pentasaccharide glycoconjugate
42a cross-reacted with the native Group A polysaccharide. Titration
curves and inhibition data further indicated that these antibodies
recognize a more extended epitope than any of the lower homologous
linear trisaccharide **31**, branched trisaccharide **33** or branched
tetrasaccharide **35**.

Monoclonal antibodies derived from mice immunized with a
killed bacterial vaccine were selected for their binding to a Group A
polysaccharide-BSA (GAP-BSA) conjugate and the BSA
glycoconjugates of the di- (*32*) and linear tri-saccharides. A promising
candidate, SA-3, showed binding to both the polysaccharide conjugate
GAP-BSA and the linear trisaccharide conjugate **40a**, and was selected
for further study.   Inhibition ELISA experiments indicated that
oligosaccharides of increasing complexity exhibited increasing potency.

Table I.   Inhibition Data[a] For Monoclonal Antibody SA-3

			Inhibitors		
Monoclonal Ab	30 μMolar	31 μMolar	33 μMolar	35 μMolar	36[b] μMolar
SA-3	97.8 (1.0)[c]	126 (1.1)	113 (1.1)	67.6 (0.8)	17.8 (0.0)

SOURCE: Adapted from Reimer, K. B.; Gidney, M. A. J.; Bundle,
D. R.; Pinto, B. M. *Carbohydr. Res.*, in press.
[a] Concentration of inhibitor compound (μMolar) required for 50%
inhibition, using solid phase antigen **40a**.
[b] This inhibitor was used as the reference for the indicated series.
[c] The values in parentheses are for $\Delta(\Delta G)$ in kcal mol^{-1}, determined
from the expression $\Delta(\Delta G)=RTln([I_1]/[I_2])$, where $[I_2]$ is the
concentration of the reference inhibitor, $[I_1]$ is the concentration of
the other inhibitors both measured at 50% inhibition; R=1.98
calK^{-1}mol^{-1}; T=295K; more positive values for $\Delta(\Delta G)$ indicate
poorer binding.

These data (Table I) suggest similar conclusions to those obtained with
the polyclonal sera raised against the branched pentasaccharide
conjugate, namely that an extended epitope is being recognized. We

suggest further that both the size and branch point are essential features of the Group A epitope.

Braun *et al.* (*33*) have previously produced monoclonal antibodies to a streptococcal group A vaccine and have observed two types of antibody: a high affinity antibody directed at an extended portion of the polysaccharide chain, with the binding site probably resembling a shallow groove, and a low affinity antibody with a small binding site that is likely to be directed at the tips of the polysaccharide chain. Extensive inhibition studies were not carried out with these antibodies, although both types were shown to bind *N*-acetylglucosamine. Our SA-3 antibody would appear to have similar binding characteristics to the high-affinity antibodies identified by Braun *et al.* (*33*).

Conclusion.   Glycoconjugates prepared from oligosaccharides of increasing complexity have been shown to provide good selectivity and discrimination in producing and characterizing carbohydrate-specific antibody. In particular, the branch point of the *Streptococcus* Group A antigen appears to be a crucial element of the epitope recognized by both polyclonal and monoclonal antibodies that are able to bind the native antigen. The synthetic methodology described above can readily furnish higher order oligosaccharides that span two or more branch points. In this regard, it is worth noting that the considerable work required to synthesize larger epitopes is justified on the basis of our preliminary immunochemical studies, a conclusion that is not always the case in responses to bacterial polysaccharides (*34*). Glycoconjugates prepared from the larger structures can facilitate the selection and characterization of antibodies with binding profiles for defined epitopes. These glycoconjugates should also provide a more accessible source of defined antigens that can be used in a variety of diagnostic formats.

Acknowledgments.   I am grateful to my colleagues who are cited in the references for their conscientious work on this project. Special thanks are due to David Bundle, my collaborator on the immunochemical aspects. I also thank the Natural Sciences and Engineering Research Council of Canada and the Heart and Stroke Foundation of B.C. and the Yukon for financial support.

Literature Cited.

1.   *Streptococcal Diseases and the Immune Response*; Read, S. E.; Zabriskie, J. B., Eds.; Academic Press: New York, N. Y., 1980.
2.   Bisno, A. L. In *Principles and Practice of Infectious Diseases, 2nd Ed.*; Mandell, G. L.; Douglas, R. G.; Bennett, J. E., Eds.; Wiley: New York, N. Y., 1985; Ch. 160, p 1133.

3.   Goldstein, I.; Halpern, B.; Robert, L. *Nature* **1967**, *213*, 44.
4.   Goldstein, I.; Rebeyrotte, P.; Parlebas, J.; Halpern, B. *Nature* **1968**, *219*, 866.
5.   Ayoub, E. M.; Shulman, S. T. In *Streptococcal Diseases and the Immune Response*; Read, S. E.; Zabriskie, J. B., Eds.; Academic Press: New York, N. Y., 1980; p 649.
6.   Goldstein, I.; Scebat, L.; Renais, J.; Hadinsky, P.; Dutartre, J. *Isr. J. Med. Sci.* **1983**, *19*, 483.
7.   Wright, K. *Science* ,**1990**, *249*, 22.
8.   Coligan, J. E.; Kindt, T. J.; Krause, R.M. *Immunochemistry* **1978**, *15*, 755.
9.   Huang, D. H.; Krishna, N. R.; Pritchard, D. G. *Carbohydr. Res.* **1986**, *155*, 193.
10. Corey, E. J.; Suggs, J. W. *J. Org. Chem.* **1973**, *38*, 3224.
11. Emery, A.; Oehlschlager, A. C.; Unrau, A. M. *Tetrahedron Lett.* **1970**, 4401.
12. Gigg, R.; Warren, C. D. *J. Chem. Soc. C* **1968**, 1903.
13. Bosshard, H. H.; Mory, R.; Schmid, M.; Zollinger, H. *Helv. Chim. Acta* **1959**, *42*, 1653.
14. Iversen, T.; Bundle, D. R. *Carbohydr. Res.* **1982**, *103*, 29.
15. Lemieux, R. U.; Abbas, S. Z.; Chung, B. Y. *Can. J. Chem.* **1982**, *60*, 58.
16. Pinto, B. M.; Morissette, D. G.; Bundle, D. R. *J. Chem. Soc. Perkin Trans. 1* **1987**, 9.
17. Hanessian, S.; Banoub, J. *Carbohydr. Res.* **1977**, *53*, 13.
18. Reimer, K. B.; Pinto, B. M. *J. Chem. Soc., Perkin Trans. 1* **1988**, 2103.
19. Reimer, K. B., Ph.D. Thesis, **1991**, Simon Fraser University, Burnaby, B.C., Canada.
20. Andrews, J. S.; Pinto, B. M. *J. Chem. Soc., Perkin Trans. 1* **1990**, 1785.
21. Andrews, J. S., M.Sc. Thesis, **1989**, Simon Fraser University, Burnaby, B.C., Canada.
22. Pinto, B. M.; Reimer, K. B.; Tixidre, A. *Carbohydr. Res.* **1991**, *210*, 199.
23. Pinto, B. M.; Buiting, M. M. W.; Reimer, K. B. *J. Org. Chem.* **1990**, *55*, 2177.
24. Peters, T.; Bundle, D. R. *Can. J. Chem.* **1989**, *67*, 491.
25. Peters, T.; Bundle, D. R. *Can. J. Chem.* **1989**, *67*, 497.
26. Kihlberg, J.; Eichler, E.; Bundle, D. R. *Carbohydr. Res.* **1991**, *211*, 59.
27. Ogawa, T.; Kitajma, T.; Nukada, T., *Carbohydr. Res.* **1983**, *123*, c5.
28. Ogawa, T.; Sugimoto, M.; Kitajma, T.; Sadozai, K. K.; Nukada, T. *Tetrahedron Lett.* **1986**, *27*, 5739.

29. Srivastava, O. P.; Hindsgaul, O. *Can. J. Chem.* **1986**, *64*, 2324.
30. Lemieux, R. U.; Bundle, D. R.; Baker, D. A. *J. Am. Chem. Soc.* **1975**, *97*, 4076.
31. Lemieux, R. U.; Baker, D. A.; Bundle, D. R. *Can. J. Biochem.* **1977**, *55*, 507.
32. Pinto, B. M.; Bundle, D. R. *Carbohydr. Res.* **1983** *124*, 313.
33. Braun, D. G., Herbst, H., Schalch, W. *The Immune System*; Karger: Basel, 1981; Vol 2.
34. Cygler, M.; Rose, D. R.; Bundle, D. R. *Science* **1991**, *253*, 442.

RECEIVED March 25, 1992

Chapter 10

Synthesis of Antigenic Heptose- and Kdo- Containing Oligosaccharides of the Inner Core Region of Lipopolysaccharides

H. Paulsen, E. C. Höffgen, and M. Brenken

Institut für Organische Chemie der Universität Hamburg, 2000 Hamburg 13, Germany

Oligosaccharides composed of Heptose (Hep) and Kdo molecules are synthesized. These compounds are partial structures of the inner core region of lipopolysaccharides. For the transformation of oligosaccharides into an immunogenic shape, two different methods are being developed. The first is coupling of the saccharide to a lipid A analogue spacer. Sheep erythrocytes can be coated with these compounds by incorporation of the fatty acids of the spacer into the cell membrane. This causes a location of the antigenic Hep and Kdo structures above the cell surface. For the second method, the allylglycosides of the corresponding oligosaccharides are required. Elongation of the allylglycosides with a cysteamine spacer under radical conditions followed by an acylation using acrylamide affords the acryloyl compounds. These are copolymerized with acrylamide. Copolymeric acrylamides are available which contains the antigenic oligosaccharides via a spacer.

Glycoconjugates and proteins located in exposed positions on the outer cell membrane of gram-negative bacteria, as illustrated in Figure 1 (*1*), are involved in important interactions between bacteria and host organism. Lipopolysaccharides (LPS) as a prominent glycoconjugate show essential physiological functions for the bacteria and exhibit a variety of biological and immunological activities in higher organisms. LPS are composed of a lipid A part anchored in the cell membrane followed by the core region and the terminal O-specific chain. The O-specific chain which determines the corresponding serotypes, shows a number of structural variations. In contrast to this, the core region of various different bacterial LPS is constructed similarly (*2*).

We have focused our interest on syntheses of fragments of the lipid A proximal inner core region, composed of the unusual sugars L-glycero-D-manno-heptose (Hep) and 3-deoxy-D-manno-octulosonic acid (Kdo). Antibodies against this inner core region should exhibit cross-reactivity or cross-protectivity against LPS of many different enterobacteria. The structure of the inner core region bound

0097–6156/93/0519–0132$06.00/0

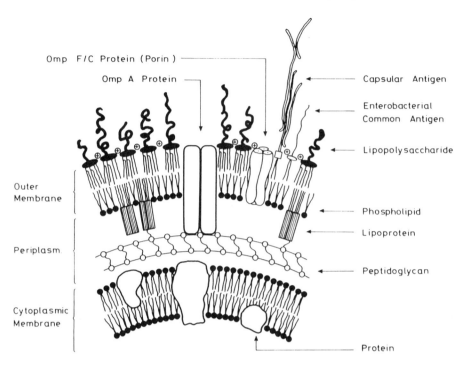

Figure 1. Cell membrane of an enterobacterial Gram-negative bacteria. (Adapted from reference 1.)

to lipid A is illustrated in formula 1. Synthetic fragments of this inner core region are required to investigate antigenic and immunogenic properties and to define, the epitope specificities of monoclonal and polyclonal antibodies (3,4).

We have synthesized a series of oligosaccharides containing Hep and Kdo. Two different methods have been developed to link these structures to polymeric

```
L-α-D -Hep
    1
    ↓
    7
L-α-D-Hep-(1→3)-L-α-D-Hep-(1→5)-α-Kdo-(2→6)-β-D-GlcN-(1→6)-α-D-GlcN
                                  4              |                   |
                                  ↑              Fatty               Fatty
                                  2             ⌊acids              acids⌋
                                α-Kdo
                                  4                   Lipid A
                                  ↑
          1                       2
                                α-Kdo
```

Chemical structure of the inner core region (dephosphorylated shape) of Salmonella LPS

carriers. The first strategy is to couple the Hep/Kdo oligosaccharides to the 6'-OH-position of disaccharide **2** (R=H) which is derivatized with two amide linked (R)-3-hydroxytetradecanoic acids (5). This structure is very similar to the natural lipid A that contains additional fatty acids and phosphates. Sheep erythrocytes can be coated with these oligosaccharides by incorporation of the fatty acids into the cell membrane through hydrophobic interactions. It has been demonstrated that these complexes carrying the antigenic Hep/Kdo determinants above the cell surface have very good immunogenic properties. The advantage of this method is, that only a small amount of valuable oligosaccharide is required (3).

R = Saccharide,Oligosaccharide

The second strategy is to copolymerize the oligosaccharide allyl glycoside with acrylamide to give linear water soluble polymers (6). The insertion of a spacer between the oligosaccharide and the polymer benefits the immunogenic properties. Therefore the allyl glycoside **3** is converted into the glycoside **4** by UV-light induced radical addition of 2-aminoethanethiol hydrochloride (cysteamine) (7). The free amino group in **4** is transformed into the acryloyl derivative **5** (8), which is copolymerized under radical conditions to afford copolymer **6** (9). The incorporation rate of the saccharide in the polyacrylamide chain is in the range of 1:15 to 1:20. The oligosaccharide in this polymer shape has also a high potential of immunogenic activity.

Kdo Oligosaccharides with a Lipid A analogous Spacer

Several Kdo oligosaccharides **7, 8, 9** coupled to the lipid A analogue Spacer **2** have been synthesized (*10,11*). The activated Kdo donor **22** is connected to a suitably protected derivative **19** with a free 6'-OH-group. The deprotection of the coupling product affords **7** containing one Kdo molecule as an immunogen. The gradual connection of a second Kdo group to a derivative of **7** furnishes **8** an

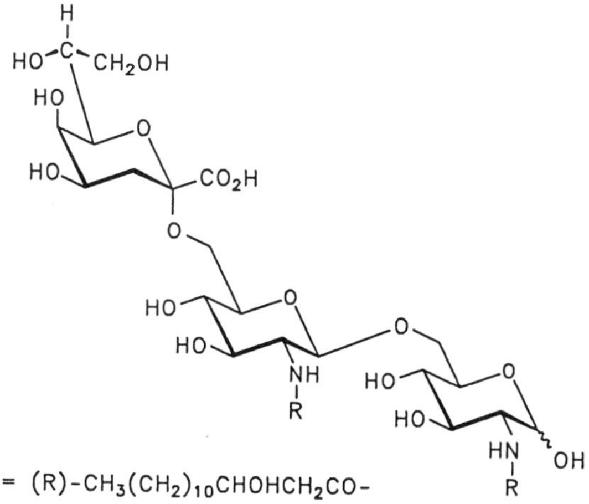

R = (R)-CH₃(CH₂)₁₀CHOHCH₂CO-

R = $(R)\text{-}CH_3(CH_2)_{10}CHOHCH_2CO\text{-}$

α-Kdo-(2→6)-β-D-GlcNhm-(1→6)-D-GlcNhm

Nhm = N-3(R)-Hydroxymyristinoyl

7

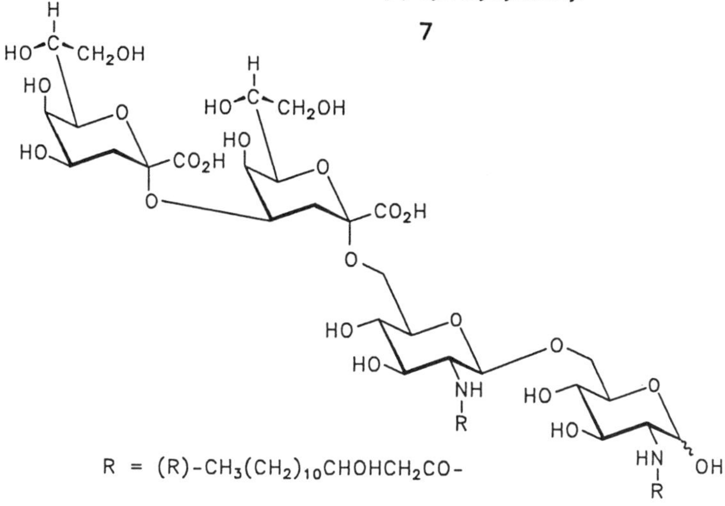

R = $(R)\text{-}CH_3(CH_2)_{10}CHOHCH_2CO\text{-}$

α-Kdo-(2→4)-α-Kdo-(2→6)-β-D-GlcNhm-(1→6)-D-GlcNhm

Nhm = N-3(R)-Hydroxymyristinoyl

8

antigenic structure with two Kdo residues. The synthesis of compound **9** containing three Kdo units requires a block synthetic strategy employing a suitable Kdo disaccharide donor that reacts with an appropriate derivative of **7**. Experimental details have been previously described (*10,11*).

Series of Kdo oligosaccharides in a polymeric shape like type **6** have been synthesized by Kosma (*12-14*). The specifity of antibodies against all of these described Kdo structures is studied by Brade (*3,4*).

$$\alpha\text{-Kdo-}(2\rightarrow 4)\text{-}\alpha\text{-Kdo-}(2\rightarrow 4)\text{-}\alpha\text{-Kdo-}(2\rightarrow 6)\text{-}\beta\text{-D-GlcNhm-}(1\rightarrow 6)\text{-D-GlcNhm}$$

Nhm = N-3(R)-Hydroxymyristinoyl

9

Kdo and Hep containing Oligosaccharides with Lipid A analogues Spacer

It is also interesting to obtain the structures containing Kdo as well as Hep in a linkage like the inner core region. The essential subject of this report is the syntheses of such type of compounds.

The central problem to be solved is the connection of a Hep unit to the very unreactive 5-OH-position of the Kdo. Therefore the 4-OH-position has to be selectively blocked as shown in derivative **11** or **12**. Benzyl ether **11** is reacted α-selectively with the activated Hep donor **13** to yield the disaccharide **14**. Gradual deprotection and following peracetylation afford **17** as a mixture with the α-anomer, which is directly transformed into the glycosyl donor **18** using titantetrabromide (TiBr$_4$). The disaccharide donor **18** is coupled with the appropriate lipid A analogue spacer **19** in the presence of mercury salt and affords the desired tetrasaccharide **20**. The yield is limited to 40% because of an inevitably occurring elimination in the Kdo part of **18**. After chromatographic

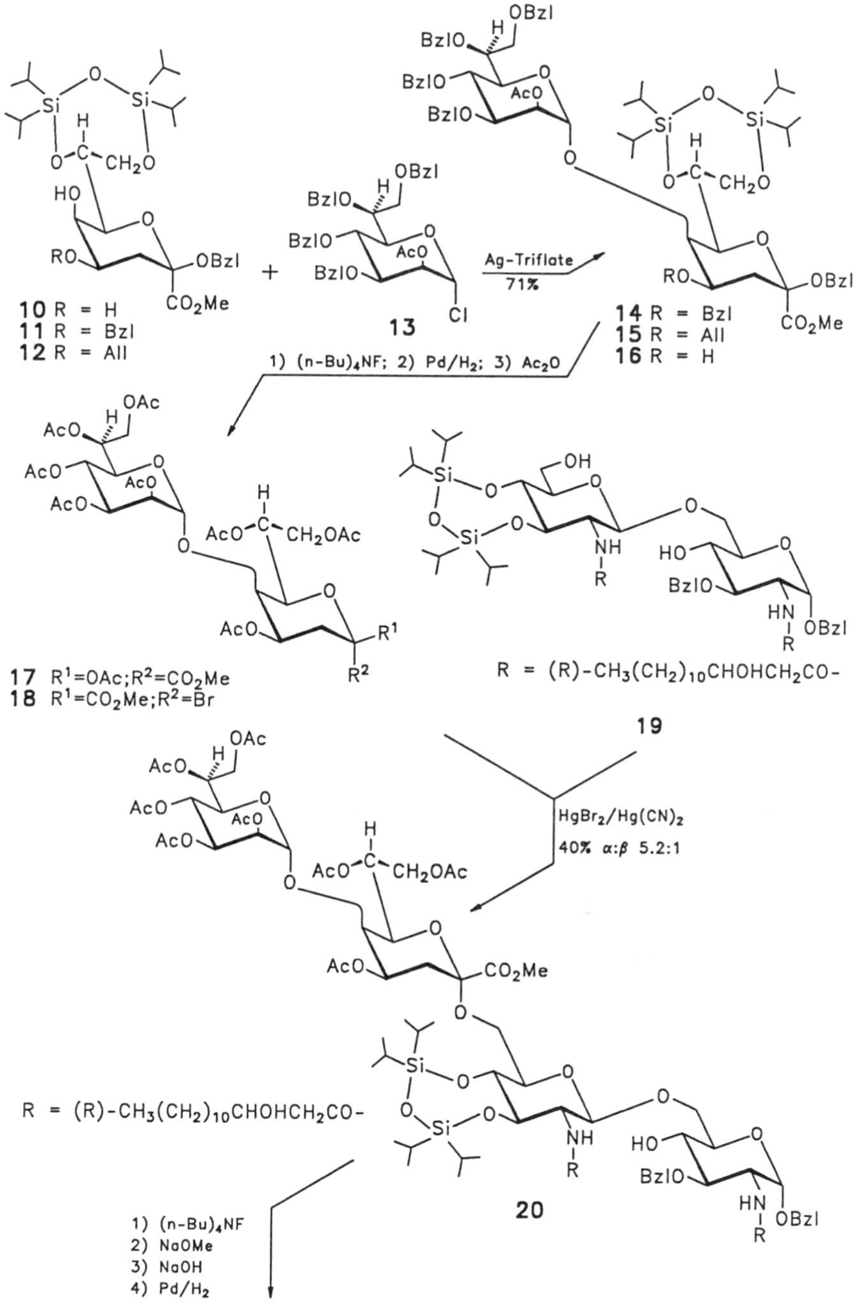

10 R = H
11 R = Bzl
12 R = All

13

14 R = Bzl
15 R = All
16 R = H

Ag-Triflate / 71%

1) (n-Bu)₄NF; 2) Pd/H₂; 3) Ac₂O

17 R¹=OAc;R²=CO₂Me
18 R¹=CO₂Me;R²=Br

R = (R)-CH₃(CH₂)₁₀CHOHCH₂CO-

19

HgBr₂/Hg(CN)₂
40% α:β 5.2:1

R = (R)-CH₃(CH₂)₁₀CHOHCH₂CO-

20

1) (n-Bu)₄NF
2) NaOMe
3) NaOH
4) Pd/H₂

L-α-D-Hep-(1→5)-α-Kdo-(2→6)-β-D-GlcNhm-(1→6)-D-GlcNhm

21

Nhm = N-3(R)-Hydroxymyristinoyl

$R = (R)-CH_3(CH_2)_{10}CHOHCH_2CO-$

L-α-D-Hep-(1→5)-α-Kdo-(2→6)-β-D-GlcNhm-(1→6)-D-GlcNhm
4
↑
2
α-Kdo

Nhm = N-3(R)-Hydroxymyristinoyl

26

separation and several deprotecting steps the pure product **21** containing Hep as well as Kdo on the lipid A analogue spacer is obtained (*15*).

Following the same strategy we have synthesized compound **23** composed of two Kdo and one Hep residues. The acceptor **16** available from **15** by cleaving the allyl ether is condensed with the bromine **22** to yield the α-linked trisaccharide **23** in an unexpectedly good yield of 80%. After removal of the tetraisopropyldisiloxane (TIPS) group with fluoride ions, hydrogenolytic cleavage of the benzyl groups, acetylation and treatment with TiBr$_4$, the obtained glycosyl donor **24** reacts with the acceptor **19** to afford the pentasaccharide **25**, in a 24%

yield which is quite good considering this complex glycosylation reaction. Step by step deblocking of **25** leads to the desired product **26** containing the antigenic trisaccharide composed of two Kdo and one Hep residues (*15*).

For the synthesis of the main chain of the inner core region consisting of two Hep and one Kdo molecules we use a different approach. An efficacious glycosyl donor of Hep, the trichloroacetimidate **27** is obtained by treatment of the known Hep hexaacetate (*16*) with hydrogenbromide/acetic acid (*17*) followed by hydrolysis of the bromide. This hydrolysis furnishes the 1-OH-free compound which reacts with trichloroacetonitrile under basic conditions to **27**. Condensation of the imidate donor **27** with the previously described Hep acceptor **28** (*18*) in the presence of trimethylsilyl-trifluoromethane sulfonate (TMS-OTf) delivered the

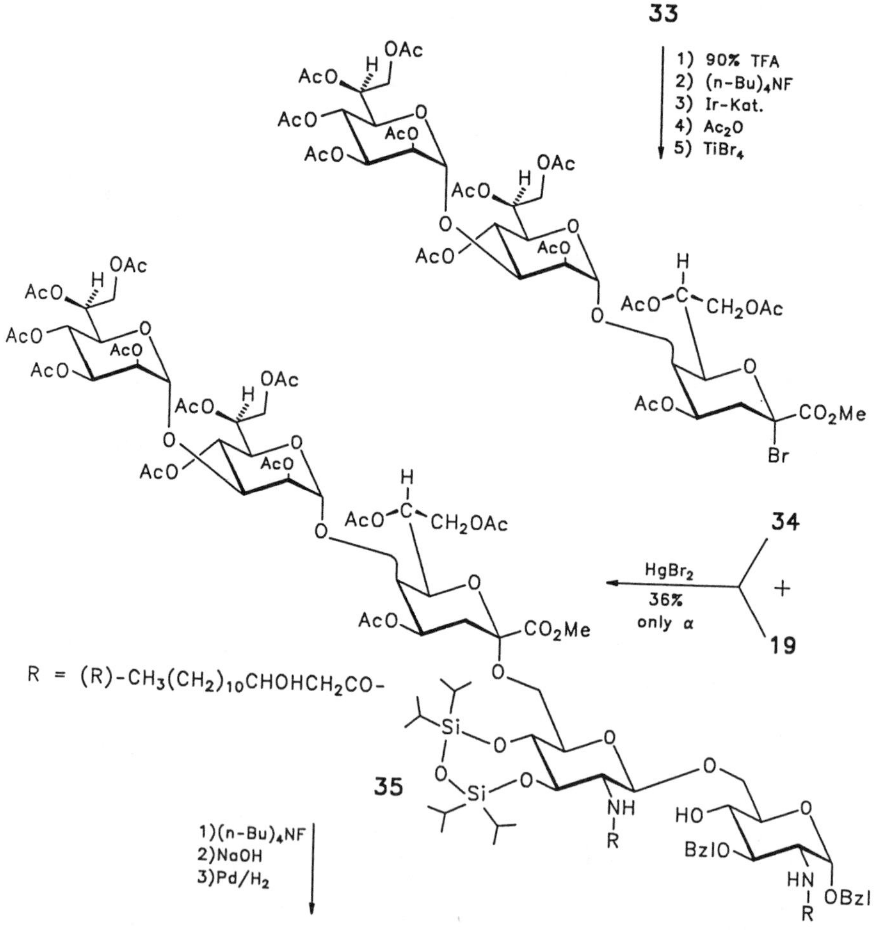

33

1) 90% TFA
2) (n-Bu)₄NF
3) Ir-Kat.
4) Ac₂O
5) TiBr₄

R = (R)-CH₃(CH₂)₁₀CHOHCH₂CO-

34

HgBr₂
36%
only α

+

19

35

1)(n-Bu)₄NF
2)NaOH
3)Pd/H₂

L-α-D-Hep-(1→3)-L-α-D-Hep-(1→5)-α-Kdo-(2→6)-β-D-GlcNhm-(1→6)-D-GlcNhm

36

Nhm = N-3(R)-Hydroxymyristinoyl

disaccharide **29** with a 96% yield. After hydrogenolysis and acetylation of **29** the α(1→3)-linked disaccharide **30** is obtainable. Hydrolysis of the 1-OH-acetyl group using piperidine and renewed treatment with trichloroacetonitrile furnishes the trichloroacetimidate **31** which is an excellent glycosyl donor as well. Glycosylation of the selectively protected Kdo acceptor **32** with the donor **31** in the presence of TMS-OTf leads stereoselectively to the desired linear trisaccharide **33** with a good yield of 72% (*19*). Initially a sequence of deblocking steps transforms **33** into the peracetylated compound which is subsequently converts to the trisaccharide donor **34** through reaction with TiBr$_4$. In spite of parallel running elimination the difficult coupling of **34** and the disaccharide spacer **19** furnishes stereoselectively to the α-linked pentasaccharide **35** with a 36% yield. Deprotection with fluoride, sodium hydroxide and subsequent hydrogenolysis delivers the deblocked pentasaccharide **36** which contains the complete main chain of the inner core region connected to a lipid A analogue spacer.

Hep and Kdo containing Oligosaccharides linked via a Spacer to Acrylamide Copolymerizates

The second method to transform oligosaccharides of the inner core region into an immunogenic shape is the coupling of saccharide structures via a spacer to acrylamide followed by a copolymerization. This strategy is demonstrated on synthesis of compound **56**. The required allyl glycoside **39** is available from **37** in two steps employing the trichloroacetimidate method. In this case it is favourable to use the benzoyl protected donor **38** instead of an acetyl protected because the intermediatly occurring orthoester **40**, can be rearranged simply with an excess of TMS-OTf into the desired glycoside **39** with a high yield. Treatment of **39** with sodium methoxide leads to compound **41**. UV-light induced addition of cysteamine followed by acylation of the amine **42** using acryloyl chloride affords **43**. The amide **43** is copolymerized with acrylamide under radical conditions to provide the linear polymer **56** which contains the saccharide structure in an incorporation rate between 1:15 to 1:20 (*20*).

$$\text{L-}\alpha\text{-D-Hep-}(1\rightarrow3)\text{-L-}\alpha\text{-D-Hep-OAll} \longrightarrow \text{Polymer } \mathbf{57}$$
46

All of the following allyl glycoside structures are transformed into corresponding copolymers **56-62** in the same way as described above. Reaction of the $\alpha(1\rightarrow3)$-linked Hep disaccharide **44** with allylalkohol furnishes the α-allyl glycoside **45**. After removal of the benzoate protecting groups, **46** is polymerized to compound **57**. Mercury salt promoted glycosylation of the

Hep/Kdo disaccharide **18** with allyl alkohol affords a mixture of the anomers with an excess of the α-anomer, that can be separated by silica gel chromatography to

yield after deprotection the pure α-anomer **47**. The pure ß-anomer **48** can be synthesized stereoselectively in a glycoside reaction under catalyst of silver silicate. Copolymerization of both compounds **47** and **48** gives the polymers **58** and **59** (*20*).

The branched trisaccharide **49** is synthesized by silver silicate promoted glycosylation of **24** with allyl alkohol. The stereoselectively formed ß-anomer is deblocked and subsequently transformed into copolymer **61**. The attempt to synthesize the corresponding α-anomer **53** employing the same bromide **24** using mercury salt catalyst results in a mixture of α- and ß-anomers which cannot be

```
        1) 90% TFA
        2) (n-Bu)₄NF
        3) NaOMe
33      4) NaOH
   ─────────────────────►   L-α-D-Hep-(1→3)-L-α-D-Hep-(1→5)-α-Kdo-OAll
                                                   54
                                                    │
                                                    ▼
                                            Polymer 62
```

seperated (*21*). For the synthesis of the pure compound **53**, a different strategy has been developed. Therefore a Kdo acceptor **32** containing a α-glycosidic allyl group from the beginning of the synthesis is used. Reaction of **27** with **32** in the presence of TMS-OTf furnishes the disaccharide **50** with good yields. Acidic cleavage of the p-methoxybenzyl group, then results in **51** which is converted with Kdo donor **22** to the branched trisaccharide **52**. After gradual deblocking to the desired pure α-allyl glycoside **53**, the copolymer **60** is obtained (*20*).

```
                CONH₂                    CONH₂
                 │                        │
  ┌┤─( CH₂-CH ─)ₓ CH₂-CH ─( CH₂-CH ─)ᵧ┤─┐
  └                        │              ┘z
                           CO
                           │
              RO〜〜S〜NH
```

```
   55  R = L-α-D-Hep —
   56  R = L-α-D-Hep-(1→7)-L-α-D-Hep—
   57  R = L-α-D-Hep-(1→3)-L-α-D-Hep—
   58  R = L-α-D-Hep-(1→5)-α-Kdo—
   59  R = L-α-D-Hep-(1→5)-β-Kdo—
   60  R = L-α-D-Hep-(1→5)-α-Kdo—
                                  4
                                  ↑
                                  2
                               α-Kdo

   61  R = L-α-D-Hep-(1→5)-β-Kdo—
                                  4
                                  ↑
                                  2
                               α-Kdo

   62  R = L-α-D-Hep-(1→3)-L-α-D-Hep-(1→5)-α-Kdo—
```

α-Allylglycoside **33** which embodies the main chain of the inner core region consisting of two Hep and one Kdo residues is synthesized as already described. Deprotection of **33** furnishes **54** followed by the preparation of the corresponding polymer **62** (*19*).

In formulae **55-62** all of the synthesized copolymers are compiled. They contain Hep and Kdo oligosaccharides as immunogenic determinants. In the first place these antigenic structures are used for preparation of antibodies against the Hep region, whereupon the study of the antibody specificities will follow.

Literature Cited

1. Rietschel, E.Th.; Brade, L.; Schade, U.; Seydel, U.; Zähringer, U.; Kusumoto, S.; Brade, H. *In Surface Structures of Microorganisms and their Interactions with the Mammalian Host*; Schrimmer, E.; Richmond, M.H.; Seibert, G.; Schwarz, U., Eds.; Workshop Conferences Hoechst; Verlag Chemie: Weinheim, 1988, Vol. 18; pp 1-14.
2. Brade, H.; Brade, L.; Rietschel, E.Th. *Zbl. Bakt. Hyg.* **1988**, *A 268*, 151-179.
3. Brade, L.; Kosma, P.; Appelmelk, B.J.; Paulsen, H.; Brade, H. *Infect. Immun.* **1987**, *55*, 462-466.
4. Rozalski, A.; Brade, L.; Kuhn, H.-M.; Brade, H.; Kosma, P.; Appelmelk, B.J.; Kusumoto, S.; Paulsen, H. *Carbohydr. Res.* **1989**, *193*, 257-270.
5. Paulsen, H.; Schüller, M. *Liebigs Ann. Chem.* **1987** 249-258.
6. Kosma, P.; Schulz, G.; Unger, F.M.; Brade, H. *Carbohydr.Res.* **1989**, *190*, 191-201.
7. Roy, R.; Tropper, F. *Glycoconjugate J.* **1988**, 203-206.
8. Roy, R.; Lafferriere, C.A. *Carbohydr. Res.* **1988**, *177*, C1-C4.
9. Norejsi, V.; Smolek, P.; Kocourek, *J. Biochem. Biophys. Act.* **1978**, *538*, 293-298.
10. Paulsen, H.; Krogmann, Ch. *Liebigs Ann. Chem.* **1989**, 1203-1213.
11. Paulsen, H.; Krogmann, Ch. *Carbohydr. Res.* **1990**, *205*, 31-44.
12. Kosma, P.; Schulz, G.; Brade, H. *Carbohydr. Res.* **1988**, *183*, 183-199.
13. Kosma, P.; Schulz, G.; Brade, H. *Carbohydr. Res.* **1989**, *190*, 191-201.
14. Kosma, P.; Bahnmüller, R.; Schulz, G.; Brade, H. *Carbohydr. Res.* **1990**, *208*, 37-50.
15. Paulsen, H.; Brenken, M. *Liebigs Ann. Chem.* **1991**, 1113-1126.
16. Paulsen, H., Schüller, M.; Heitmann, A.; Nashed, M.A.; Redlich, H. *Liebigs Ann. Chem.* **1986** 675-686.
17. Boons, G.J.P.H.; Overhand, M.; von der Marel, G.A.; van Boom, J.H. *Carbohydr. Res.* **1989**, *192*, C1-C4.
18. Paulsen, H.; Heitmann, A.C. *Liebigs Ann. Chem.* **1988**, 1061-1071.
19. Paulsen, H.; Höffgen, E.C. *Tetrahedron Lett.* **1991**, *32*, 2747-2750.
20. Paulsen, H.; Brenken, M.; Wulff, A. *Liebigs Ann. Chem.*, **1991**, 1127-1145
21. Paulsen, H.; Brenken, M.; Krogmann, Ch.; Heitmann, A.C.; Wulff, A. *In Cellular and Molecular Aspects of Endotoxin Reaction*; Nowotny, A.; Spitzer, J.J.; Ziegler, E.J., Eds.; Endotoxin Research Series, Experta Medica; Elsevier: Amsterdam-New York-Oxford, 1990, Vol. 1; pp 51-60

RECEIVED April 24, 1992

Chapter 11

Molecular Biology of Anti-α-(1→6)dextrans

Antibody Responses to a Single-Site-Filling Antigenic Determinant

Elvin A. Kabat

Department of Microbiology, College of Physicians and Surgeons, Columbia University, New York, NY 10032

α(1→6)dextrans are homopolymers of glucose with chains of α(1→6)linked glucoses generally connected by short branches with α(1→2), α(1→3) and α(1→4)linkages (1). During the 1980's extensive improvements in determining dextran structures have been made in analytical periodate structural analysis (POSA), methylation structural analysis (MSA), non-reducing end group analysis (NREG) which have made possible more exact correlations with immunochemical studies and led to an improved notation (1) as well as to the elimination of ambiguities resulting from the use of (1→6)-like instead of (1→) for a terminal non-reducing end and (1→6) for two possibilities. The second residue may be α(1→2)-, α(1→3)-, α(1→4)-, or α(1→6)-linked in various dextrans (cf 1). Dextran B512F which was generally given as having 95% α(1→6) and 5% α(1→3)linkages actually has 90% α(1→6), 5% α(1→) and 5% α(1→3;b) with b denoting a branch point. Oligosaccharide chains containing 2-10 or more α(1→6)linked glucose oligosaccharides have been isolated (2,3). α(1→6)linked dextrans with molecular weights of over 90,000 have been shown to be antigenic in humans (4) giving rise to precipitating antibodies and to wheal and erythema type skin sensitivity which seriously restrict its use as a plasma volume expander. Partially hydrolyzed fractions of dextran have been shown to be less antigenic and products of molecular weights

0097–6156/93/0519–0146$06.00/0

of about 50,000 or below did not induce antibody
responses in humans (4).

α(1→6)dextran   itself   and   α(1→6)linked
oligosaccharides of dextran coupled to stearylamine to
give synthetic glycolipids are both T-independent
antigens in mice (5-9) giving rise to precipitating
monoclonal hybridoma antibodies of the IgM and IgA
variety plus several IgG3 hybridomas (10). α(1→6)linked
oligosaccharides coupled to BSA and KLH have been found
to be T dependent antigens in mice (11). Moreover, the
studies on mouse monoclonal myeloma proteins initiated
by Michael Potter (5), have augmented the available
repertoire of anti-α(1→6) specific monoclonals and have
made it desirable to use these to study the scope and
nature of the repertoire of hybridoma antibodies to a
well defined and characterized single epitope - a chain
of α(1→6)linked glucoses which fill the combining sites
of various myeloma and hybridoma anti-α(1→6)dextrans.
It must be emphasized that the methodology developed for
these studies is immunochemical and is unrelated to and
independent of x-ray crystallographic methods of
exploring antibody combining sites.

Size of Anti-α(1→6)dextran combining sites
     The size of anti-α(1→6)dextran combining sites is
defined by the largest oligosaccharide which will
saturate the site so that larger oligosaccharides are
equal in inhibiting activity on a molar basis (12, cf
13). This is a quite reproducible finding with
monoclonals but is not seen with heterogeneous
populations of antibodies produced by direct immunization
with dextran. Figure 1 shows the result of immunization
of humans with two injections of 0.5 mg of native
α(1→6)dextran a day apart (13). This was done before
such studies were prohibited. These data are of great
interest. The six individuals of whom I was subject No.
1, since I always believed that I should inject myself
with any material before it was used on other persons,
show decidedly different patterns of anti-α(1→6)dextran
responses. Thus comparing the relative quantities of
isomaltotriose   (IM3),   isomaltotetraose   (IM4),
isomaltopentaose (IM5), isomaltohexaose (IM6) giving 50
percent inhibition of precipitation, it can be seen that
the ratios of IM6:IM5:IM4:IM3 differed from one human
antiserum to another. Thus the ratio of IM6:IM3 for the
six sera studied varied from 30:1 to 2.5:1. For each
antiserum the same amount of antibody and the same
concentration of inhibitor on a molar basis were
compared. Thus the inhibition data were directly
comparable in terms of molecules. These studies were
carried out by quantitative precipitin inhibition assays
using the quantitative precipitin methodology developed

by Michael Heidelberger and Forrest E. Kendall in 1929 (15).

Were one dealing with identical populations of antibodies with respect to site size and binding constant, such as is seen with myeloma proteins or hybridomas, the six sets of curves should have been superimposable. Thus Figure 1 establishes that each of the six antisera were polyclonal mixtures of anti-α(1→6)dextrans with combining sites of different sizes and of different Ka.

Subsequently, Schlossman (16) and Gelzer (17) fractionated my anti-α(1→6)dextran into portions with predominantly smaller and larger size sites by absorbing substantial quantities (ca 2 liter samples) of my serum on Sephadex G25, washing out extraneous antibodies and serum proteins and eluting antibodies with smaller size sites with IM3 and then larger size sites with IM6. They confirmed that the antiserum contained mixtures of anti-α(1→6)dextrans with smaller and larger sized combining sites as had been inferred from Figure 1.

The next important development came from two monoclonal myeloma cell lines, W3129 (18,20) and QUPC52 (19) which secreted anti-α(1→6)dextrans. These were studied by Cisar et al. (20,21). Measurement of site sizes of these two antibodies revealed important differences in site structure. W3129 had a site saturated by isomaltopentaose, whereas that of QUPC52 accommodated isomaltohexaose. Their binding constants were $1x10^5$ for W3129 and $5x10^3$ for QUPC52. Since previous studies had shown that an antibody with a larger site size had a higher Ka, we measured the relative contribution of each glucose to the total binding by inhibition using equilibrium dialysis.

The result was extraordinary. For W3129 with the smaller site, methyl α-D-glucoside and isomaltose (IM) contributed 50 percent of the total binding of the IM5 which saturated the site, whereas with QUP52 with the larger site, methyl α-D-glucoside and isomaltose contributed less than five percent of the total binding of the site-filling IM6 and only with IM3 did significant binding occur. The structural basis of these findings became clear when Ruckel and Schuerch (22) generously provided a synthetic linear α(1→6)dextran with 200 glucoses. With W3129 it inhibited precipitation by dextran on a molar basis equivalent to IM5. With QUPC52, the striking finding was that it precipitated the anti-α(1→6)dextran.

These results indicated that the W3129 site was

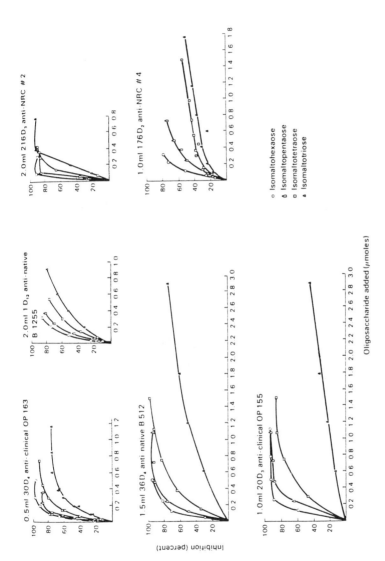

Figure 1. Inhibition by isomaltose oligosaccharides of precipitation by dextran of the antidextran in six human antisera. (From Kabat and Mayer's Experimental Immunochemistry. (2d ed.; 1961) Charles C Thomas, Publisher, Springfield, ILL.)

directed toward the non-reducing terminal glucose plus
the four glucoses to which it was joined.   QUPC52
precipitated because the linear dextran was multivalent
and contained many epitopes of six α(1→6)glucoses not
including the reducing end.   These findings were
interpreted as indicating that W3129 had a cavity-type
site (21) (end binder (23)) into which the non-reducing
glucose could fit with the contiguous four glucoses in
a groove.   QUPC52, however, had a groove-type site into
which a chain of six glucoses could fit not involving the
non-reducing end.   This has proven quite advantageous in
screening anti-α(1→6)hybridoma secretions for groove or
cavity type sites by doing a precipitin test on the
culture supernatants.   Bennett and Glaudemans (24)
confirmed the cavity-type site in W3129.

The distinction between groove-type and cavity-type
sites was verified when Borden et al (25) made anti-
idiotypic sera and showed that anti-cavity type sites
reacted only with the idiotype to cavity-type (W3129)
anti-α(1→6)dextran and the anti-idiotype recognizing
groove-type sites reacted only with groove-type sites
and when tested with a large number of groove-type anti-
dextrans showed different degrees of reactivity
indicating that it was polyclonal recognizing a number
of distinct idiotypes.   Thus anti-idiotypic sera can play
a role in recognizing aspects of the diversity of the
repertoire of antibodies to a site-filling epitope.

The distinction between groove and cavity type sites
(21,22) based entirely on immunochemical evidence (21)
made it independent of x-ray crystallography and
naturally led to attempts to model the two kinds of
antibody combining sites.   The first models were of
cavity-type W3129 and of groove-type 19.1.2 (26, cf 27)
FV domains.   The V_H and V_L of W3129 (28) and of 19.1.2
(29) had been sequenced in the laboratory.   The atomic
coordinates of the V-domains used were from the crystal
structures of the phosphocholine binding Fab complex (30)
refined at 3.1A° resolution and those of J539 were from
T. N. Bhat (31).   Models of the Fv were built by the
interactive graphics program FRODO (32) adapted to an
Evans and Sutherland PS300 system.   Each domain was built
assuming that the individual domains had the same
backbone structures.   Differences in length occurred only
in the CDRs and these were chosen based on similarities
in sequence and length.   For V_H W3129 the starting model
was that of J539 to which it was almost identical even
in V_H CDR1 and V_H CDR2.   In V_H CDR3 W3129 was closer to
McPC603 which was used with the insertion of one residue
after Ser 99. V_L of W3129 was somewhat more similar to
that of McPC603 than to that of J539 with both having Pro
95 in the cis position. V_H McPC603 was used with the

exision of residue 27f from McPC603 and the ends
rejoined. The amino acid sequence of V_L 19.1.2 resembled
J539 must closely having V_L Pro 95 and was used for all
but the last five residues which were from McPC603.
V_H19.1.2 was also more like J539 but was one residue
shorter in CDR3 and one residue longer in J539.
Accordingly J539 was used in the starting model with V_H
CDR3 of J539 with residue 100 excised the two models were
then energy minimized.

The modeled combining sites for W3129 and 19.1.2 and
their surface representations may be found in (26, see
27). W3129 and 19.1.2 differ extensively in the lengths
of the CDRs and in their amino acids. A wall formed by
the long V_L CDR1 and V_H CDR3 protrudes on one side of the
combining site which is not seen in 19.1.2 because of the
much shorter V_L CDR1. V_L CDR1 is one of the shortest
known. The combining site surface proposed for 19.1.2
is relatively flat and is seen as a shallow groove
whereas that proposed for W3129 has a cavity in the
middle resulting from the smaller residues Gly and Leu
at positions 91 and 96 respectively in V_LCDR3 which in
19.1.2 are the bulky Tyr and Trp which prevent formation
of a cavity like that seen in the model of W3129. Both
the W3129 and 19.1.2 models can accommodate several sugar
residues.

Model building is considered a very high risk
enterprise. However, the proposed models of W3129 and
19.1.2 are based on quantitative immunochemical data and
thus have direct evidence of combining site structure,
not available to earlier model building studies.

Additional structural data on anticarbohydrate sites
was developed by Glaudemans (33) by the use of deoxy and
deoxyfluoro-oligosaccharides which permitted examination
of hydrogen bonding patterns. Thus replacement of a
hydroxyl group of a hapten by a hydrogen will eliminate
its contribution to binding. Use of a fluorinated
oligosaccharide would prevent hydrogen donation by the
oligosaccharide. Such changes in binding are generally
measured by fluorescence quenching (or enhancement). The
C-F bonds the van der Waals ratio of the C-F bond 1.39
and $1.37A°$ are shorter than those for a hydrogen bond C-
O, 1.43 and $1.41A°$, so that spatial effects would
essentially be negligible. These additional measurements
were obtained by Nashed et al (34) for both W3129 and a
second cavity-type site, 16.4.12E, an anti-α(1→6)dextran
obtained by immunizing with IM6-KLH (35).

Table I summarizes the binding constants Ka obtained
with oligosaccharides, deoxyoligosaccharides, and
fluorodeoxy oligo- saccharides as well as the maximum

Table I. Comparison of K_a and maximum ligand-induced fluorescence change (ΔF_{max}) for glucose derivatives with the IgAs 16.4.12E and W3129

Ligand[a]	16.4.12E		W3129	
	K_a	ΔF_{max}	K_a	ΔF_{max}
	M^{-1}	(%)	M^{-1}	(%)
1 Glc→αMe	4.5×10^3	$(+51)^b$	1.8×10^3	$(-19)^c$
2 Glc$_2$→αMe	5.2×10^4	$(+45)^b$	1.6×10^4	$(-16)^c$
3 Glc$_3$→αMe	8.7×10^4	$(+41)^b$	6.7×10^4	$(-16)^c$
4 Glc$_4$→αMe	3.9×10^5	$(+47)^b$	1.8×10^5	$(-13)^c$
5 Glc$_5$→αMe	2.4×10^5	$(+47)^b$	1.9×10^5	$(-14)^c$
6 Glc$_6$→αMe	3.8×10^5	$(+45)^b$	1.8×10^5	$(-15)^c$
7 2dGlc→αMe	0.2×10^2	$(+25)^d$	2.9×10^2	$(-12)^d$
8 2FGlc→αMe	0.8×10^2	$(+36)^d$	1.8×10^3	$(-15)^d$
9 6FGlcα(1→2)Glc→αMe	0^e		0	
10 6FGlcβ(1→2)Glc→αMe	0		0	
11 3dGlc→αMe	0		0.3×10^2	$(-47)^d$
12 3FGlc→αMe	1.2×10^2	$(+36)^d$	7.7×10^3	$(-15)^d$
13 6FGlcα(1→3)Glc→αMe	0		0	
14 6FGlcβ(1→3)Glc→αMe	0		2.1×10^2	$(-18)^c$
15 Galβ(1→3)Glc→αMe	0		1.8×10^2	$(-16)^d$
16 4dGlc→αMe	0		0	
17 4FGlc→αMe	0		0	
18 6dGlc→αMe	0		0	
19 6FGlc→αMe	0.3×10^2	$(+47)^d$	0	
20 6-O-MeGlc→αMe	0.6×10^2	$(+50)^d$	0	
21 6FGlc→Glc→αMe	3.6×10^2	$(+40)^b$	0.5×10^2	$(-12)^c$
22 6FGlc→Glc$_2$→αMe	6.3×10^2	$(+48)^b$	2.6×10^2	$(-16)^c$
23 6FGlc→Glc$_3$→αMe	3.7×10^3	$(+49)^b$	4.6×10^2	$(-19)^c$
24 Gal→βMe	0		0	

[a]Glc→αMe, methyl α-D-glucopyranoside; Glc$_2$→αMe, methyl O-(α-glucopyranosyl)-(1→6)α-D-glucopyranoside; Glc$_3$→α, the corresponding trisaccharide-glycoside, etc. (unless specifically noted, for instance as in **9**, all linkages are α(1→6), 2dGlc→αMe, methyl 2-deoxy-α-glucopyranoside; 2FGlc→αMe, the corresponding 2-deoxy-2-fluoro glycoside, etc.
[b]Values for the correlation factor (R^2) are between 0.998 and 0.984.
[c]From Glaudemans et al. (33)
[d]Values for the correlation factor (R^2) are between 0.983 and 0.895.
[e]By values of 0 we mean that the K_a is below quantifiable level by this method.
From (34).

induced fluorescent change, ΔF max (34). The two
antibodies clearly have cavity-type sites. In each case
the nonreducing end contributes the maximum induced
fluorescence change; compound 1 and the higher glycosides
compounds 2-6, show that almost the same tryptophanyl
fluorescence change is induced by the ligand. Me α-D-
isomaltotetraose binds optimally indicating that it may
fill the site. However measurements with free isomaltose
oligosaccharides indicate that the CH2 of the fifth
glucose from the non-reducing end contributes some
additional binding energy. Methyl 6-deoxy-αD-
glucopyranoside or methyl 6-deoxy-6-fluoro-α-D-glucoside
do not bind indicate that the H of the 6-hydroxy group
of the ligand must donate a H in binding to the combining
site. The only sugar in the epitope with a 6-OH group
is the non-reducing glucose. Thus subsite (A) of the
antibody (Figure 2) binds the terminal non-reducing end,
a finding in agreement with Matsuda and Kabat (35) and
the four subsites are A, B, C and D with binding
constants of 4.5 X 10^3, 11.6, 1.7 and 4.5 respectively.
The four possible methyl monodeoxy glucopyranosides
7,11,16,18 showed nearly complete loss of binding while
the 3-deoxyfluoro, 12, and 2-deoxyfluoro, 8, showed only
very weak binding. The 4- and 6- deoxyfluoroglycosides
show almost complete loss of binding. Thus the 4- and
6-OH groups are involved in H-bond donation while the 3-
and 2-OH groups show weak shared H bond reactions. The
role of these 4- and 6-OH thus determine the nonreducing
glucose as the immunodominant group of the epitope.
Lemieux et al. (36) previously reported that an α-
galactosyl non-reducing terminal group of anti-I Ma was
also immunodominant.

The findings with W3129 and 16.4.12E are very similar
and the results of model building are of great interest.
Thus W3129 had earlier been shown to have a cavity-type
site into which a glucose could fit by its non reducing
end filling the cavity with the rest of the site being
a groove. However, 16.4.12E had a substantially larger
site somewhat displaced from that of W3129 and it was
necessary to add two molecules of water to fill the site
and position the non-reducing glucose. The importance
of water was earlier emphasized by Quicho and Vyas (37)
and Vyas (38) for a carbohydrate binding protein.
Colored stereo models of the two sites may be found in
(34).

The proposed model of W3129 shows the only solvent
exposed Trp residues to be in the wall of a deep cavity
near the H-L interface in the combining area of the Fv
domain. Me αDGlc affects the Trp fluorescence to the
same extent as all other ligands which bind its cavity
and thus is proposed to be the major affinity subsite of
all ligands that bind.

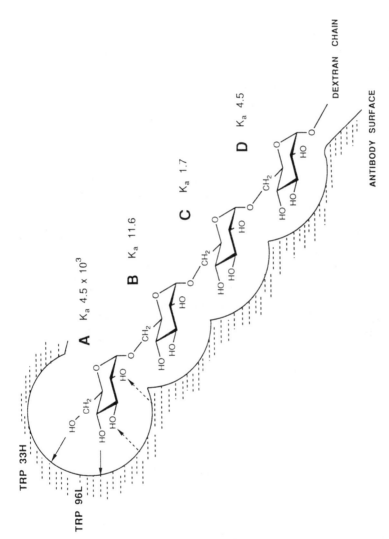

Figure 2. Schematic representation of the subsites of IgA 16.4.12E, showing the tetrasaccharide epitope and the individual affinity constants each glucosyl residue has for its subsite. From (34).

16.4.12E has a more intense Trp fluorescence at positions 33H, 100AH and 96L. Two Trp residues come together on the solvent exposed wall of a pronounced cavity at the H/L interface of the Ig in the same general location as in W3129. The fluorescence change for all six oligosaccharide glycosides is the same within experimental error indicating that the major affinity subsite is in the cavity. The placement in the cavity of methyl αDglucoside was such that the 4- and 6-OH groups were able to interact with the carboxyl groups of Gly 96H and 98H and were at 3.1 and 2.8A° respectively while the OH groups of positions 2 and 3 were placed near (3.1 and 3.2A°) a water molecule in the cavity H bonded to O(3.4A°) of Glu 50H.

The finding that both W3129 and 16.4.12E show cavities of different sizes and that a non reducing glucose fills the cavity of W3129 but that 16.4.12E has a somewhat larger cavity requiring two H bonded water molecules suggests that there may be a population of cavity-type molecules of different sizes. 16.4.12E perhaps because of its larger size site (39) than W3129 does not show the same idiotypic specificity. X ray studies of these types of sites as well as of groove-type anti-α(1→6)dextran sites may shed new light on the repertoire of antibody combining sites directed toward a single antigenic determinant (40).

Wang et al (40) have examined the number of $(V_H D J_H) \times (V_L J_L)$ combinations of sequenced anti-α(1→6) monoclonal hybridoma and myeloma proteins and have defined 18 combinations, 15 with groove-type and three with cavity-type sites. 11 V_H genes of 7 V_H subgroups (I, IIA, IIB, IIIC, and V for groove-type sites, IB and IIIB for cavity-type sites); 7 V_κ genes of subgroups (II, III, V and VI for the groove-type, II for the cavity-type); 9 D minigenes and all active J_H minigenes are used. The sets of diverse mAb were further expanded by junctional diversity, N element insertions, D-D fusion and somatic point mutation (40). P insertions and gene conversions are not yet observed in anti-α(1→6)dextrans. The population of antibody combining sites to the single antigenic determinant become quite large and comparable those seen to other more complicated antigen systems.

The finding that in forming anti-α(1→6)dextrans, so many germ-line genes are used to generate similar-type combining sites indicates that the antibody forming system is highly protected against catastrophic germ line gene loss since even if one or more of the nine germ line gene families were to disappear the individual could still make a reasonably complete repertoire of anti-$\alpha(1\rightarrow6)$dextran combining sites.

The laboratory has also (41) identified two major families of monoclonal anti-$\alpha(1\rightarrow6)$dextrans, with prototypes designated $V_H19.1.2$ and $V_H9.14.7$ which show distinct patterns of J_κ and J_H minigene usage with different amino acid substitutions in CDR3 all have groove-type sites; there are V_H10 19.1.2 and 11 $V_H9.14.7$ mAb. Both families use the same D minigene (DFL16.1). 19.1.2 used only J_H2 and J_H3 whereas 9.14.7 used J_H1, J_H2, and J_H3 and all four active J_κ, $J_\kappa1$, $J_\kappa2$, $J_\kappa4$, $J_\kappa5$). All used the V_κ-Ox1 germ-line gene (Ref. 21 and 22). These data reveal some of the influences of genes and minigenes in controlling the types of antibody combining sites to a single epitope. The nature of these influences requires further study and may well play a dominant role in the generation of the other carbohydrate and non carbohydrate determinants.

Acknowledgments

The author thanks Drs. Denong Wang and Niel Glaudemans for going over the manuscript.

Aided by grants from the National Institute of Allergy and Infectious Diseases, 1RO1 AI-19042, 5RO1 AI-125616, 3RO1 AI-127508 and the National Science Foundation DMB 8600778 to EAK of Columbia University, and the National Cancer Institute NCI-CA13696 to Columbia University.

Work with the PROPHET computer system is supported by a grant to Columbia University from the National Institute of Allergy and Infectious Diseases, the National Cancer Institute, the National Institute of Diabetes, Digestive and Kidney Diseases, the National Institute of General Medical Sciences and the National Library of Medicine with a subcontract from Columbia University to Bolt Beranek and Newman.

References

1. Jeanes, A.   Mol. Immunol.   23:999–1028, 1986 and papers of F. Seymour cited therein. See also for an improved notation for representing dextran structures which had led to improved understanding of anti-$\alpha(1\rightarrow6)$dextrans and their combining sites. See also Rees, D.A., Richardson, N.G., Wright, N.J. and Hirst,

E.   Carb. Res. 9:451-462, 1969 for distinguishing between linear and branched oligosaccharides.

2. Turvey, J.R. and Whelan, W.J. Biochem. J. 67:49-52, 1957.

3. Lai, E. and Kabat, E.A.   Mol. Immunol. 22:1021-1037, 1985.

4. Kabat, E.A. and Bezer, A.E.   Arch Biochem. Biophys. 78:306-318, 1958.

5. Potter, M.   Physiol Revs. 52:631-719, 1972.

6. Sharon, J., Kabat, E.A. and Morrison, S.L.   Mol. Immunol. 18:831-846, 1981.

7. Wood, C. and Kabat, E.A.   J. Exp. Med. 154:432-449, 1981.

8. Wood, C. and Kabat, E.A.   Arch. Biochem. Biophys. 212:262-276; 277-289, 1981.

9. Lai, E., Kabat, E.A. and Mobraaten, L.   Cellular Immunol. 92:172-183, 1985.

10. Chen, H.T., Makover, S.D. and Kabat, E.A.   Mol. Immunol. 24:333-338, 1987.

11. Matsuda, T. and Kabat, E.A.   J. Immunol. 142:863-870, 1989.

12. Kabat, E.A.   J. Immunol. 84:82-85, 1960.

13. Kabat, E.A.   Structural Concepts in Immunology and Immunochemistry, Second Edition Holt, Rinehart and Winston, 1976.

14. Kabat, E.A.   J. Immunol., 97:1-11, 1966.

15. Heidelberger, M. and Kendall, F.E.   J. Exp. Med. 50:809-823, 1929.

16. Schlossman, S.F. and Kabat, E.A.   J. Exp. Med. 116:535-552, 1962.

17. Gelzer, J. and Kabat, E.A.   Immunochemistry 1:303-316, 1964.

18. Weigert, M., Raschke, W.C., Carson, D. and Cohn, M. J. Exp. Med. 139:127-146, 1974.

19. QUPC52: Obtained from the NIH collection of Michael Potter.

20. Cisar, J.O., Kabat, E.A., Liao, J. and Potter, M. J. Exp.   Med. 139:159-179, 1974.

21. Cisar, J., Kabat, E.A., Dorner, M.M. and Liao, J. J. Exp. Med. 142:435-459, 1975.

22. Ruckel, E.R. and Schuerch, C.   J. Am. Chem. Soc. 88:2605-2606, 1966.

23. Davies, D.R. and Metzger, H.   Ann. Rev. Immunol. 1:87-117, 1983.

24. Bennett, L. and Glaudemans, C.P.J.   Carbohy. Res. 72:315-319, 1979.

25. Borden, P. and Kabat, E.A. Mol. Immunol. 25:251-262, 1988.

26. Padlan, E.A. and Kabat, E.A.   Proc. Natl. Acad. Sci. 85:6885-6889, 1988.

27. Padlan, E.A. and Kabat, E.A.   Methods in Enzymology, Academic Press NY   203:3-21, 1991.

28. Borden, P. and Kabat, E.A.   Proc. Natl. Acad. Sci. 84:2440-2443, 1987.

29. Sikder, S.K., Akolkar, P.N., Kaladas, P.M., Morrison, S.L. and Kabat, E.A. J. Immunol. 135:4215-4221, 1985

30. Padlan, E.A., Cohen, G.H. and Davies, D.R.   Protein
    Data Bank, File No. 2, MCP.
31. Bhat, T.N., Padlan, E.A. and Davies, D.R.   Protein
    Data Bank, File 2FBJ.
32. Jones, T.A.   J. Applied Crystallography 11:268-272,
    1978.
33. Glaudemans, C.P.J.   Chem. Rev. 91:25-33, 1991.
34. Nashed, E.M., Perdomo, G.R., Padlan, E.A., Kovac, P.,
    Matsuda, T., Kabat, E.A. and Glaudemans, C.P.J.   J.
    Biol. Chem. 265:20699-20707, 1990.
35. Matsuda, T. and Kabat, E.A.   J. Immunol. 142:863-870,
    1989.
36. Lemieux, R.U., Wang, T.C., Liao, J. and Kabat, E.A.
    Mol. Immunol. 21:751-759, 1989.
37. Quicho, F.A. and Vyas, N.K.   Nature 310:381-386, 1984
38. Vyas, N.K., Vyas, M.N. and Quicho, F.A.   Science
    242:1290-1295, 1988.
39. Wang, D. and Kabat, E.A.   unpublished.
40. Wang, D., Liao, J., Mitra, D., Akolkar, P.N., Gruezo,
    F. and Kabat, E.A.   Mol. Immunol. 28:1387-1397,1991.
41. Wang, D., Chen, H-T., Liao, J., Akolkar, P.N.,
    Sikder, S., Gruezo, F. and Kabat, E.A.   J. Immunol.
    145:3002-3010, 1990.
42. Kabat, E.A.   The Physico-Chemical Biology (Japan)
    34:11-24, 1990.

RECEIVED August 10, 1992

Chapter 12

Extended Type 1 Chain Glycosphingolipids

Le^a–Le^a (Dimeric Le^a) and Le^b–Le^a as Human Tumor
Associated Antigens

Mark R. Stroud, Steven B. Levery, and Sen-itiroh Hakomori

**The Biomembrane Institute, 201 Elliott Avenue West,
Seattle, WA 98119**

Aberrant glycosylation associated with animal and human cancers has been
clearly defined by improved chemical analysis of carbohydrates (CHOs) and
introduction of monoclonal antibody (MAb) technology in tumor immunology. While ganglio-series and globo-series antigens have been found to be
highly expressed as tumor-associated antigens (TAAs) in specific types of
human cancer such as Burkitt lymphoma, melanoma, and neuroblastoma,
lacto-series structures are more widely expressed in such common human
cancers as gastrointestinal, pancreatic, lung, breast, and hepatic cancers (see
for review *1*). Two types of lacto-series antigens are known: type 1 and
type 2 chain. Their fucosylated forms were discovered in 1964, and their
structures were characterized subsequently (*2-4*). It was noticed that Le^a
and Le^b antigens were both accumulated in human cancers regardless of host
Lewis status. However, they were not widely recognized as human TAAs
until the MAb approach was introduced in human tumor immunology.
Accumulation of glycosphingolipids (GSLs) having type 2 chain with fucosylated or sialosylated/fucosylated structure (particularly extended form) in a
variety of human cancers has been well-documented (*5-8*). Type 1 chain
lacto-series antigens have long been considered to exist only as non-extended,
non-repeated structures (*9*; see for review *1*). Recently, MAb NCC-ST-421
(hereafter termed "ST-421") was established using primary human gastric
cancer ST-4 inoculated subcutaneously as xenograft in Balb/c-Nu/nu mice.
The xenograft acted as immunogen, but immunization was performed by
transferring immunocompetent normal Balb/c mouse spleen cells into host
Balb/c-Nu/nu mice bearing ST-4 tumor xenografts (*10*). When xenograft
tumors regressed, spleen cells of host mice were fused with HAT-sensitive
myeloma, and hybridomas were screened for preferential reactivity with
tumors on formalin-fixed paraffin sections, as described previously for
establishment of other anti-TAA MAbs (*11*). The antigen defined by MAb

0097–6156/93/0519–0159$06.00/0
© 1993 American Chemical Society

ST-421 was identified as Le^a/Le^a, i.e., $\alpha1{\to}4$ fucosylated extended type 1 chain (Table I, structure 1). Subsequently, another extended type 1 chain antigen with Le^b/Le^a structure (struc. 2) was isolated and characterized from Colo-205 cells, and MAb IMH2 directed against this antigen was established. Isolation and chemical characterization of the Le^a/Le^a and Le^b/Le^a antigens, and immunobiological properties of their respective MAbs ST-421 and IMH2, will be summarized in this paper.

Isolation and Characterization of Le^a/Le^a (Dimeric Le^a; $IV^4FucV^4FucLc_6$)

This antigen was found during the search for GSL compounds reacting with MAb ST-421, which was known to have a strong inhibitory effect on growth of human tumors in nude mice. Isolation of Le^a/Le^a from pooled human colon cancer extracts and from human colon cancer cell line Colo205 has been described previously (12). The GSL fraction was extracted and puri-fied, and the slow-migrating ST-421-reactive band was separated by prepara-tive thin-layer chromatography (TLC). Structural determination was per-formed by enzymatic degradation, positive-ion fast atom bombardment mass spectrometry ($^+$FAB-MS), methylation analysis, and proton nuclear magnetic resonance spectroscopy (^{1}H-NMR). Results clearly indicated that the antigen has extended type 1 chain structure with $\alpha1{\to}4$ fucosylation at the penultimate as well as internal GlcNAc. The ^{1}H-NMR spectrum and $^+$FAB-MS spectrum of permethylated GSLs are shown in Figures 1 and 2 respec-tively (12).

Binding Specificity of MAb ST-421

ST-421 reacts strongly with Le^a/Le^a antigen. Binding specificity of this MAb is shown in Figure 3. It reacts equally well with synthetic GSL antigen Le^a/Le^x (Table I, struc. 5). This antigen has not been detected as a glyco-lipid; however, it is possible that Le^a/Le^x occurs as a side chain on glycopro-teins.

Isolation and Characterization of Le^b/Le^a ($III^4FucV^4FucVI^2FucLc_6$)

This antigen was found as another slow-migrating GSL component of Colo205 cell extract, and was purified on ion-exchange chromatography, high-pressure liquid chromatography, and preparative TLC. ^{1}H-NMR and $^+$FAB-MS clearly indicate the structure of this antigen to be extended type 1 chain with $\alpha1{\to}2$ fucosylation at the terminal Gal and $\alpha1{\to}4$ fucosylation at the penultimate GlcNAc as well as internal GlcNAc (Table I, struc. 2) (13). The ^{1}H-NMR spectrum and $^+$FAB-MS spectrum of this antigen are shown in Figures 4 and 5 respectively.

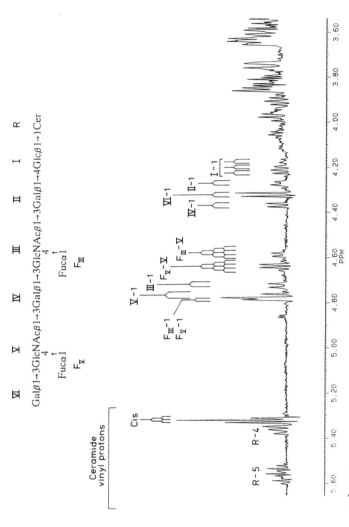

Figure 1. ¹H-NMR spectrum of Le^a/Lc^a antigen isolated from Colo205 cells (downfield region). Arabic numerals refer to ring protons of residues designated by roman numerals in the corresponding structure shown at top of figure. R refers to protons of the sphingosine backbone only; Cis refers to vinyl protons of unsaturated fatty acids. Fuc H-5/CH₃ connectivities confirmed by decoupling. (Reproduced with permission from reference 12. Copyright 1991 The American Society for Biochemistry & Molecular Biology.)

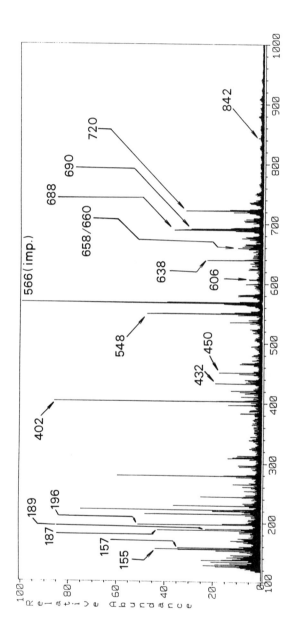

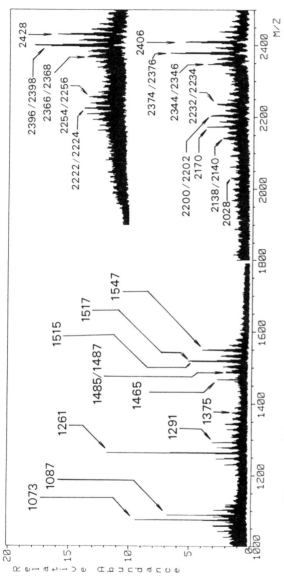

Figure 2. ⁺FAB mass spectrum of permethylated Le^a/Le^a. Composite of three acquisitions optimized for sensitivity under different conditions. Segment from 100-1800 a.m.u. was acquired with NBA only as matrix. Lower segment from 1800-2500 a.m.u. was acquired with addition of 15-Crown-5 to matrix. Insert segment (1900-2500 a.m.u.) was acquired with addition of sodium acetate to matrix. All assignments are nominal monoisotopic masses.
(Reproduced with permission from reference 12. Copyright 1991 The American Society for Biochemistry & Molecular Biology.)

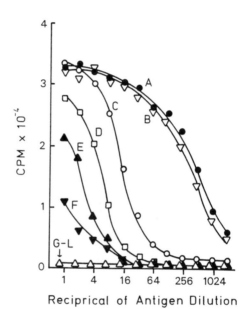

Figure 3. Reactivity of Le^a/Le^a, Le^a/Le^x, and related GSLs with MAb ST-421. Binding activity of ST-421 with GSLs coated on solid phase was determined as previously described (12). A, Le^a/Le^x (Table I, structure 5). B, Le^a/Le^a (struc. 1). C, extended Le^a without internal $\alpha 1 \rightarrow 4$ fucose (struc. 8). D, Le^a pentasaccharide ceramide (III^{4}FucLc$_4$) (struc. 9). E, IV^{3}Galβ1→3GlcNAcIII^{3}FucnLc$_4$ (struc. 10). F, Le^x/Le^x (struc. 4). G, type 1 chain H (IV^{2}FucLc$_4$). H, IV^{3}Galβ1→3GlcNAcnLc$_4$ (struc. 11). I, Le^b hexasaccharide ceramide (struc. 7). J, nLc$_6$ (struc. 12). K, VI^{2}FucnLc$_6$ (H$_2$ antigen). L, trifucosyl Le^y (struc. 3).

(Reproduced with permission from reference 12. Copyright 1991 The American Society for Biochemistry & Molecular Biology.)

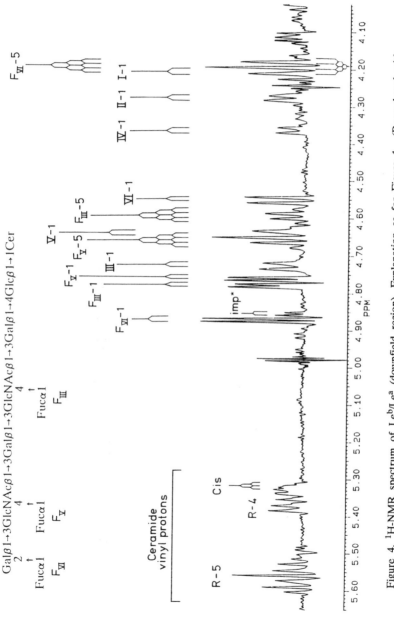

Figure 4. ^{1}H-NMR spectrum of Le^b/Le^a (downfield region). Explanation as for Figure 1. (Reproduced with permission from reference 24. Copyright 1992 American Association for Cancer Research, Inc.)

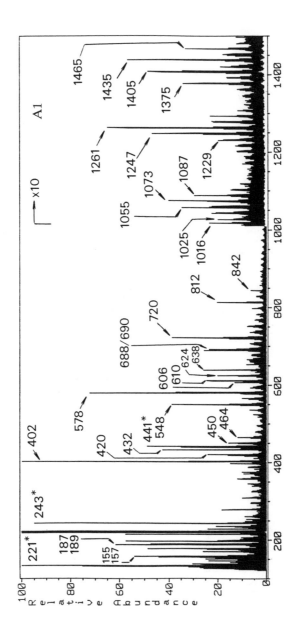

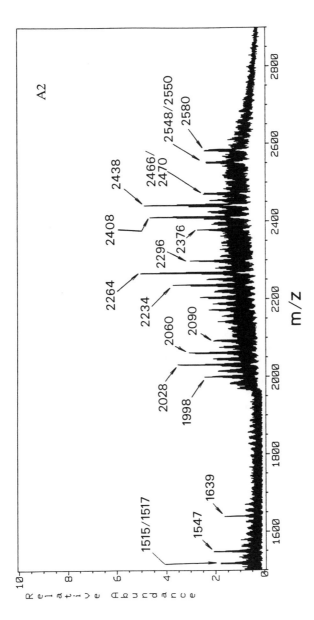

Figure 5A. +FAB mass spectrum of permethylated Le^b/Le^a. Matrix: NBA/15-crown-5. All assignments are nominal, monoisotopic masses. (Reproduced with permission from reference 13. Copyright 1992 Springer.)

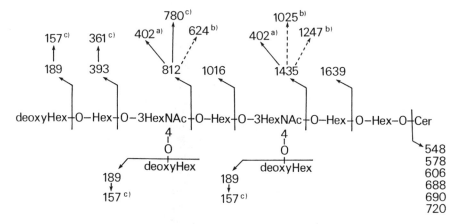

Figure 5B. Proposed fragmentaion scheme of permethylated Le^b/Le^a. Matrix: NBA/15-crown-5. All assignments are nominal, monoisotopic masses. (Reproduced with permission from reference 13. Copyright 1992 Springer.)

Table I. Structures and trivial names of extended and fucosylated type 1 and type 2 chain antigens

1.
$$Gal\beta1\rightarrow3GlcNAc\beta1\rightarrow3Gal\beta1\rightarrow3GlcNAc\beta1\rightarrow3Gal\beta1\rightarrow4Glc\beta1\rightarrow1Cer$$
$$4 4$$
$$\uparrow \uparrow$$
$$Fuc\alpha1 Fuc\alpha1 Le^a/Le^a$$

2.
$$Gal\beta1\rightarrow3GlcNAc\beta1\rightarrow3Gal\beta1\rightarrow3GlcNAc\beta1\rightarrow3Gal\beta1\rightarrow4Glc\beta1\rightarrow1Cer$$
$$2 4 4$$
$$\uparrow \uparrow \uparrow$$
$$Fuc\alpha1 Fuc\alpha1 Fuc\alpha1 Le^b/Le^a$$

3.
$$Gal\beta1\rightarrow4GlcNAc\beta1\rightarrow3Gal\beta1\rightarrow3GlcNAc\beta1\rightarrow3Gal\beta1\rightarrow4Glc\beta1\rightarrow1Cer$$
$$2 3 3$$
$$\uparrow \uparrow \uparrow$$
$$Fuc\alpha1 Fuc\alpha1 Fuc\alpha1 \text{trifucosyl } Le^y$$

4.
$$Gal\beta1\rightarrow4GlcNAc\beta1\rightarrow3Gal\beta1\rightarrow4GlcNAc\beta1\rightarrow3Gal\beta1\rightarrow4Glc\beta1\rightarrow1Cer$$
$$3 3$$
$$\uparrow \uparrow$$
$$Fuc\alpha1 Fuc\alpha1 Le^x/Le^x$$

5.
$$Gal\beta1\rightarrow3GlcNAc\beta1\rightarrow3Gal\beta1\rightarrow4GlcNAc\beta1\rightarrow3Gal\beta1\rightarrow4Glc\beta1\rightarrow1Cer$$
$$4 3$$
$$\uparrow \uparrow$$
$$Fuc\alpha1 Fuc\alpha1 Le^a/Le^x$$

6.
$$Gal\beta1\rightarrow4GlcNAc\beta1\rightarrow3Gal\beta1\rightarrow4Glc\beta1\rightarrow1Cer$$
$$2 3$$
$$\uparrow \uparrow$$
$$Fuc\alpha1 Fuc\alpha1 Le^y \text{ hexasaccharide ceramide}$$

7.
$$Gal\beta1\rightarrow3GlcNAc\beta1\rightarrow3Gal\beta1\rightarrow4Glc\beta1\rightarrow1Cer$$
$$2 4$$
$$\uparrow \uparrow$$
$$Fuc\alpha1 Fuc\alpha1 Le^b \text{ hexasaccharide ceramide}$$

8.
$$Gal\beta1\rightarrow3GlcNAc\beta1\rightarrow3Gal\beta1\rightarrow4GlcNAc\beta1\rightarrow3Gal\beta1\rightarrow4Glc\beta1\rightarrow1Cer$$
$$4$$
$$\uparrow$$
$$Fuc\alpha1 \text{extended } Le^a$$

9.
$$Gal\beta1\rightarrow3GlcNAc\beta1\rightarrow3Gal\beta1\rightarrow4Glc\beta1\rightarrow1Cer$$
$$4$$
$$\uparrow$$
$$Fuc\alpha1 Le^a \text{ pentasaccharide ceramide}$$

10.
$$Gal\beta1\rightarrow3GlcNAc\beta1\rightarrow3Gal\beta1\rightarrow4GlcNAc\beta1\rightarrow3Gal\beta1\rightarrow4Glc\beta1\rightarrow1Cer$$
$$3$$
$$\uparrow$$
$$Fuc\alpha1$$

11.
$$Gal\beta1\rightarrow3GlcNAc\beta1\rightarrow3Gal\beta1\rightarrow4GlcNAc\beta1\rightarrow3Gal\beta1\rightarrow4Glc\beta1\rightarrow1Cer$$

12.
$$Gal\beta1\rightarrow4GlcNAc\beta1\rightarrow3Gal\beta1\rightarrow4 nLc_6$$

Establishment of MAb IMH2, and its Antigen-Binding Specificity

MAB IMH2 was established after immunization of Balb/c mice with *Salmonella minnesotae* coated with Le^b/Le^a antigen, followed by screening of hybridomas based on reactivity with various related GSL antigens, as previously described (*14*). We obtained hybridoma IMH2, which secreted an IgG_3 MAb reacting strongly with not only Le^b/Le^a antigen but also Le^y/Le^x (trifucosyl Le^y) (Table I, struc. 3). The MAb reacted to a lesser extent with Le^y and Le^b hexasaccharide ceramides (struc. 6 and 7). GSLs with Le^a/Le^a, Le^a, Le^x, or Le^x/Le^x reacted very weakly or not at all (Ito H, Stroud MR, Nudelman ED, unpubl.) (See Figure 6.)

Immunobiological Properties of MAbs ST-421 and IMH2

ST-421 showed strong growth-inhibitory effect on various antigen-positive human tumors grown in nude mice. However, it did not inhibit growth of antigen-negative tumors (i.e., tumors not expressing Le^a/Le^a). Since the antibody isotype of ST-421 is IgG_3, it showed strong antibody-dependent cellular cytotoxicity (ADCC) and complement-dependent cytotoxicity (CDC) against antigen-positive tumors in the presence of human lymphocytes and complement (*10*). *In vivo* tumor-suppressive effect of ST-421 against various carcinoma cell lines (*10*) is shown in Figure 7.

IMH2 showed ADCC and CDC against antigen-positive tumor cells (e.g. Colo205), and inhibited tumor (Colo205) growth in nude mice, to a similar degree as ST-421 (Figure 8).

Cell Biological and Functional Significance of TAAs in General

Why are Le^a/Le^a, Le^b/Le^a, and other fucosylated or sialosylated/fucosylated structures predominantly expressed in a wide variety of human carcinomas? Despite many years of study, the answer to this question remains a mystery. These structures are strong antigens, and presumably have important immunobiological roles, as suggested in this paper. On the other hand, many of them also function as adhesion molecules, or modifiers of adhesive proteins or their receptors (e.g., integrin receptors). Le^x-expressing cells adhere to other Le^x-expressing cells or matrix, based on homotypic Le^x/Le^x interaction (*15*). Similarly, H-expressing cells adhere to Le^y-expressing cells, based on H/Le^y interaction (*16*). These observations were based on cell adhesion occurring in embryogenesis. However, there is also evidence that CHO-CHO interactions are involved in initial recognition and adhesion of tumor cells to microvascular endothelial cells (ECs), leading to metastatic deposition. Sialosyl-Le^x (SLe^x) and sialosyl-Le^a (SLe^a) are major TAAs (*1*), and have recently been identified as the epitopes recognized by the selectins GMP-140 and ELAM-1, which are expressed on activated platelets and ECs (*17-21*). The ability of tumor cells to activate platelets and ECs, and subse-

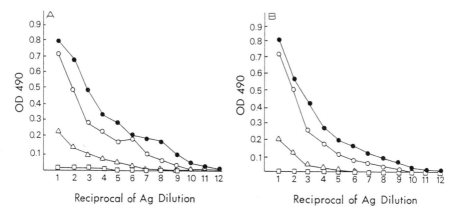

Figure 6. Reactivity of MAb IMH2 with various GSLs. Serial double dilutions of various GSL antigens mixed with cholesterol and lectin in a molar ratio of 1:5:3 were added to 96-well flat-bottom assay plates (Probind plate, Falcon) in ethanol and dried. Initial concentration of GSL added to the first well was 100 ng/well. MAb binding assay was performed by ELISA as described in the text. Abscissa, reciprocal of antigen dilution. Ordinate, optical density reading at 490 nm.

Panel A: reactivity of type 1 chain GSLs. ●, Le^b/Le^a. ○, Le^b hexasaccharide ($IV^2FucIII^4FucLc_4$). △, Le^a/Le^a □ (all of the following showed similar reactivity), type 1 chain PG (Lc_4); H_1 type 1 chain (IV^2FucLc_4); Le^a/Le^x ($IV^3Gal\beta1{\to}3[Fuc\alpha1{\to}4]GlcNAc\beta III^3FucnLc_4$).

Panel B: reactivity of type 2 chain GSLs. ●, trifucosyl Le^y ($VI^2FucV^3FucIII^3FucnLc_6$). ○, Le^y hexasaccharide ceramide ($IV^2FucIII^3$-$FucnLc_4$). △, Le^x/Le^x ($V^3FucIII^3FucnLc_6$). □ (all of the following had similar reactivity), H_1 type 2 chain (IV^2nLc_4); H_2 type 2 chain (IV^2Fuc-Lc_6); PG (nLc_4); Le^x ($III^3FucnLc_4$).

(Reproduced with permission from reference 13. Copyright 1992 Springer.)

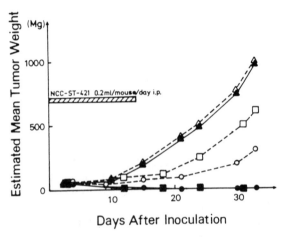

Days After Inoculation

Figure 7. Inhibition of tumor growth by MAb ST-421 *in vivo*. Human tumor xenografts (2-3 mm diameter) were inoculated subcutaneously, and experimental mice (solid lines) received intraperitoneal administration for 2 weeks of ST-421 in ascites (0.2 ml). Control animals (dashed lines) from each group were treated with ascites produced by mouse myeloma cell line P3-X63-Ag8-U1. Human gastric tumor line St-4 (●, ○) and human colonic tumor line Co-4 (■, □) were positive for the antigen defined by ST-421, while human breast tumor line MX-1 (▲, △) was negative for the antigen. Solid and hollow symbols represent experimental and control animals, respectively.

(Reproduced with permission from reference 10. Copyright 1991 American Association for Cancer Research, Inc.)

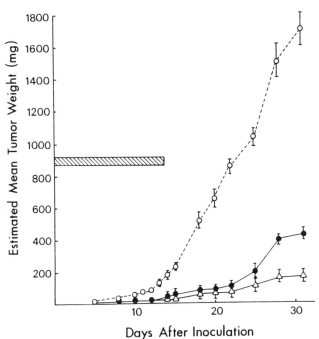

Figure 8. Inhibitory effect of MAb IMH2 on Colo205 cell growth in nude mice. 1x10^7 Colo205 cells were s.c. injected into the backs of 6-wk-old athymic Balb/c mice, followed immediately by injection of 200 μl ($\approx$200 μg) of purified IMH2 (1.1 mg/ml) per day for 14 days (shaded bar) (●). Other mice were treated similarly with MAb ST-421 (△). Control groups were injected with PBS containing similar quantities of non-specific mouse IgG (○). (Reproduced with permission from reference 24. Copyright 1992 American Association for Cancer Research, Inc.)

quently adhere to them, is a crucial factor in initiation of metastasis. Therefore, expression of SLe^x, SLe^a, or other sialosylated/fucosylated structures on tumor cells may be instrumental in tumor progression.

There is also recent evidence that fucosylated structures (e.g., H/Le^b/-Le^y) may help control tumor cell motility. MAb MIA-15-5, directed to H/Le^b/Le^y, selected based on inhibitory effect on tumor cell motility, was found to inhibit not only motility but also metastatic potential of tumor cells (22). It is possible that ST-421 and IMH2 also inhibit tumor cell motility, although such studies have not yet been performed. ST-421 and IMH2 both inhibited growth of human tumors in nude mice (see preceding section), and showed strong ADCC and CDC in the presence of human lymphocytes and complement, but did not show strong ADCC or CDC in the presence of mouse lymphocytes and complement. This unusual finding suggests that the tumor-suppressive effect of these MAbs is based on inhibition of tumor cell motility or adhesion. Indeed, ST-421 was found to block adhesion of antigen-positive cells.

There is still no direct evidence that fucosylated or sialosylated/fucosylated structures are involved in human tumor progression. However, expression of these structures in primary human tumors has been negatively correlated with patient survival in several studies (see for review 23). It seems entirely possible, therefore, that therapeutic application of MAbs directed to these TAAs, or of the CHO structures themselves, could block human tumor progression.

Acknowledgments

We thank Dr. Stephen Anderson for scientific editing and preparation of the manuscript. Studies described from the authors' laboratory were supported by National Cancer Institute Outstanding Investigator Grant CA42505 (to S.H.) and funds from The Biomembrane Institute, in part under a research contract with Otsuka Pharmaceutical Co.

Literature Cited

1.   Hakomori, S. *Adv. Cancer Res.* **1989**, *52*, 257-331.
2.   Hakomori, S.; Jeanloz, R.W. *J. Biol. Chem.* **1964**, *239*, 3606-3607.
3.   Yang, H.-J.; Hakomori, S. *J. Biol. Chem.* **1971**, *246*, 1192-1200.
4.   Hakomori, S.; Andrews, H. *Biochim. Biophys. Acta* **1970**, *202*, 225-228.
5.   Hakomori, S.; Nudelman, E.D.; Levery, S.B.; Kannagi, R. *J. Biol. Chem.* **1984**, *259*, 4672-4680.
6.   Fukushi, Y.; Hakomori, S.; Nudelman, E.D.; Cochran, N. *J. Biol. Chem.* **1984**, *259*, 4681-4685.
7.   Fukushi, Y.; Nudelman, E.D.; Levery, S.B.; Rauvala, H.; Hakomori, S. *J. Biol. Chem.* **1984**, *259*, 10511-10517.

8.   Kaizu, T.; Levery, S.B.; Nudelman, E.D.; Stenkamp, R.E.; Hakomori, S. *J. Biol. Chem.* **1986**, *261*, 11254-11258.

9.   Kannagi, R.; Levery, S.B.; Hakomori, S. *J. Biol. Chem.* **1985**, *260*, 6410-6415.

10. Watanabe, M.; Ohishi, T.; Kuzuoka, M.; Nudelman, E.D.; Stroud, M.R.; Kubota, T.; Kodaira, S.; Abe, O.; Hirohashi, S.; Shimosato, Y.; Hakomori, S. *Cancer Res.* **1991**, *51*, 2199-2204.

11. Watanabe, M.; Hirohashi, S.; Shimosato, Y.; Ino, Y.; Yamada, T.; Teshima, S.; Sekine, T.; Abe, O. *Jpn. J. Cancer Res. (Gann)* **1985**, *76*, 43-52.

12. Stroud, M.R.; Levery, S.B.; Nudelman, E.D.; Salyan, M.E.K.; Towell, J.A.; Roberts, C.E.; Watanabe, M.; Hakomori, S. *J. Biol. Chem.* **1991**, *266*, 8439-8446.

13. Stroud, M.R.; Levery, S.B.; Salyan, M.E.K.; Roberts, C.E.; Hakomori, S. *Eur. J. Biochem.* **1992**, *203*, 577-586.

14. Young, W.W.Jr.; MacDonald, E.M.S.; Nowinski, R.C.; Hakomori, S. *J. Exp. Med.* **1979**, *150*, 1008-1019.

15. Eggens, I.; Fenderson, B.A.; Toyokuni, T.; Dean, B.; Stroud, M.R.; Hakomori, S. *J. Biol. Chem.* **1989**, *264*, 9476-9484.

16. Fenderson, B.A.; Kojima, N.; Stroud, M.R.; Zhu, Z.; Hakomori, S. *Glycoconjugate J.* **1991**, *8*, 179.

17. Phillips, M.L.; Nudelman, E.D.; Gaeta, F.C.A.; Perez, M.; Singhal, A.K.; Hakomori, S.; Paulson, J.C. *Science* **1990**, *250*, 1130-1132.

18. Polley, M.J.; Phillips, M.L.; Wayner, E.A.; Nudelman, E.D.; Singhal, A.K.; Hakomori, S.; Paulson, J.C. *Proc. Natl. Acad. Sci. USA* **1991**, *88*, 6224-6228.

19. Berg, E.L.; Robinson, M.K.; Mansson, O.; Butcher, E.C.; Magnani, J.L. *J. Biol. Chem.* **1991**, *266*, 14869-14872.

20. Takada, A.; Ohmori, K.; Takahashi, N.; Tsuyuoka, K.; Yago, A.; Zenita, K.; Hasegawa, A.; Kannagi, R. *Biochem. Biophys. Res. Commun.* **1991**, *179*, 713-719.

21. Handa, K.; Nudelman, E.D.; Stroud, M.R.; Shiozawa, T.; Hakomori, S. *Biochem. Biophys. Res. Commun.* **1991**, *181*, 1223-1230.

22. Miyake, M.; Hakomori, S. *Biochemistry* **1991**, *30*, 3328-3334.

23. Hakomori, S. *Curr. Opin. Immunol.* **1991**, *3*, 646-653.

24. Ito, H.; Tashiro, K.; Stroud, M.R.; Ørntoft, T.F.; Meldgaard, P.; Singhal, A.K.; Hakomori, S. *Cancer Res.* **1992**, *52*, 3739–3745.

RECEIVED April 14, 1992

INDEXES

Author Index

Affiliation Index

Subject Index

179

Production: Margaret J. Brown
Indexing: Deborah H. Steiner
Acquisition: Barbara C. Tansill
Cover design: Alan I. Kahan

Printed and bound by Maple Press, York, PA

Highlights from ACS Books

Good Laboratory Practices: An Agrochemical Perspective
Edited by Willa Y. Garner and Maureen S. Barge
ACS Symposium Series No. 369; 168 pp; clothbound, ISBN 0–8412–1480–8

Silent Spring Revisited
Edited by Gino J. Marco, Robert M. Hollingworth, and William Durham
214 pp; clothbound, ISBN 0–8412–0980–4; paperback, ISBN 0–8412–0981–2

Insecticides of Plant Origin
Edited by J. T. Arnason, B. J. R. Philogène, and Peter Morand
ACS Symposium Series No. 387; 214 pp; clothbound, ISBN 0–8412–1569–3

Chemistry and Crime: From Sherlock Holmes to Today's Courtroom
Edited by Samuel M. Gerber
135 pp; clothbound, ISBN 0–8412–0784–4; paperback, ISBN 0–8412–0785–2

Handbook of Chemical Property Estimation Methods
By Warren J. Lyman, William F. Reehl, and David H. Rosenblatt
960 pp; clothbound, ISBN 0–8412–1761–0

The Beilstein Online Database: Implementation, Content, and Retrieval
Edited by Stephen R. Heller
ACS Symposium Series No. 436; 168 pp; clothbound, ISBN 0–8412–1862–5

Materials for Nonlinear Optics: Chemical Perspectives
Edited by Seth R. Marder, John E. Sohn, and Galen D. Stucky
ACS Symposium Series No. 455; 750 pp; clothbound; ISBN 0–8412–1939–7

Polymer Characterization:
Physical Property, Spectroscopic, and Chromatographic Methods
Edited by Clara D. Craver and Theodore Provder
Advances in Chemistry No. 227; 512 pp; clothbound, ISBN 0–8412–1651–7

From Caveman to Chemist: Circumstances and Achievements
By Hugh W. Salzberg
300 pp; clothbound, ISBN 0–8412–1786–6; paperback, ISBN 0–8412–1787–4

The Green Flame: Surviving Government Secrecy
By Andrew Dequasie
300 pp; clothbound, ISBN 0–8412–1857–9

For further information and a free catalog of ACS books, contact:
American Chemical Society
Distribution Office, Department 225
1155 16th Street, NW, Washington, DC 20036
Telephone 800–227–5558

Bestsellers from ACS Books

The ACS Style Guide: A Manual for Authors and Editors
Edited by Janet S. Dodd
264 pp; clothbound, ISBN 0–8412–0917–0; paperback, ISBN 0–8412–0943–X

Chemical Activities and Chemical Activities: Teacher Edition
By Christie L. Borgford and Lee R. Summerlin
330 pp; spiralbound, ISBN 0–8412–1417–4; teacher ed. ISBN 0–8412–1416–6

Chemical Demonstrations: A Sourcebook for Teachers,
Volumes 1 and 2, Second Edition
Volume 1 by Lee R. Summerlin and James L. Ealy, Jr.;
Vol. 1, 198 pp; spiralbound, ISBN 0–8412–1481–6;
Volume 2 by Lee R. Summerlin, Christie L. Borgford, and Julie B. Ealy
Vol. 2, 234 pp; spiralbound, ISBN 0–8412–1535–9

Writing the Laboratory Notebook
By Howard M. Kanare
145 pp; clothbound, ISBN 0–8412–0906–5; paperback, ISBN 0–8412–0933–2

Developing a Chemical Hygiene Plan
By Jay A. Young, Warren K. Kingsley, and George H. Wahl, Jr.
paperback, ISBN 0–8412–1876–5

Introduction to Microwave Sample Preparation: Theory and Practice
Edited by H. M. Kingston and Lois B. Jassie
263 pp; clothbound, ISBN 0–8412–1450–6

Principles of Environmental Sampling
Edited by Lawrence H. Keith
ACS Professional Reference Book; 458 pp;
clothbound; ISBN 0–8412–1173–6; paperback, ISBN 0–8412–1437–9

Biotechnology and Materials Science: Chemistry for the Future
Edited by Mary L. Good (Jacqueline K. Barton, Associate Editor)
135 pp; clothbound, ISBN 0–8412–1472–7; paperback, ISBN 0–8412–1473–5

Personal Computers for Scientists: A Byte at a Time
By Glenn I. Ouchi
276 pp; clothbound, ISBN 0–8412–1000–4; paperback, ISBN 0–8412–1001–2

Polymers in Aqueous Media: Performance Through Association
Edited by J. Edward Glass
Advances in Chemistry Series 223; 575 pp;
clothbound, ISBN 0–8412–1548–0

For further information and a free catalog of ACS books, contact:
American Chemical Society
Distribution Office, Department 225
1155 16th Street, NW, Washington, DC 20036
Telephone 800–227–5558